CW00523908

Molecular Biology for Obstetricians and Gynaecologists

Phillip Bennett
BSc MD MRCOG

Gudrun Moore
BA PhD

The Action Research Laboratory for
the Molecular Biology of Fetal Development,
Royal Postgraduate Medical School,
Institute of Obstetrics and Gynaecology,
Queen Charlotte's and Chelsea Hospital,
London

ILLUSTRATIONS BY
Philip Stainier
BSc PhD

FOREWORD BY
Charles Rodeck
DSc FRCOG
Professor of Obstetrics and Gynaecology
University College & Middlesex
School of Medicine,
University of London

OXFORD
BLACKWELL SCIENTIFIC PUBLICATIONS
LONDON EDINBURGH BOSTON
MELBOURNE PARIS BERLIN VIENNA

Editorial Offices:
Osney Mead, Oxford OX2 0EL
25 John Street, London WC1N 2BL
23 Ainslie Place, Edinburgh EH3 6AJ
3 Cambridge Center, Cambridge
Massachusetts 02142, USA
54 University Street, Carlton
Victoria 3053, Australia

Other Editorial Offices:
Librairie Arnette SA
2, rue Casimir-Delavigne
75006 Paris
France

Blackwell Wissenschafts-Verlag
Meinekestrasse 4
D-1000 Berlin 15
Germany

Blackwell MZV
Feldgasse 13
A-1238 Wien
Austria

First published 1992

Set by Setrite Typesetters, Hong Kong
Printed and bound in Great Britain by
The Alden Press, Oxford

DISTRIBUTORS

Marston Book Services Ltd
PO Box 87
Oxford OX2 0DT
(*Orders*: Tel: 0865 791155
Fax: 0865 791927
Telex: 837515)

USA
Blackwell Scientific Publications, Inc.
3 Cambridge Center
Cambridge, MA 02142
(*Orders*: Tel: 800 759–6102)

Canada
Times Mirror Professional
Publishing, Ltd
5240 Finch Avenue East
Scarborough, Ontario M1S 5AZ
(*Orders*: Tel: 416 298–1588)

Australia
Blackwell Scientific Publications
(Australia) Pty Ltd
54 University Street
Carlton, Victoria 3053
(*Orders*: Tel: 03 347–0300)

British Library
Cataloguing in Publication Data

Bennett, Phillip
Molecular biology for obstetricians and gynaecologists.
I. Title II. Moore, Gudrun
574.88

ISBN 0–632–02744–4

Contents

Part 2: Molecular Biology Techniques

Part 3: The Clinical Applications of Molecular Biology

Foreword

Facts accumulate with the advances of science, but, as Medawar pointed out in *The Art of the Soluble*, as science matures, the facts become subordinated to and incorporated into generalizations, laws or principles of increasing explanatory power. This has been the impact that molecular biology has had on the biological sciences and the shock waves are spreading into all areas of clinical medicine. Within obstetrics and gynaecology, prenatal diagnosis and screening, oncology, endocrinology and maternal and fetal viral infections are all being transformed. Together with new knowledge and new diagnostic and therapeutic procedures comes a new language, which clinicians must learn if they are not to remain illiterate.

This book has grown out of the workshops held by the authors at the Institute of Obstetrics and Gynaecology at Queen Charlotte's and Chelsea Hospital. It is clearly written and illustrated, has an essential glossary and is not cluttered with references, but gives a useful guide to further reading. The authors themselves (an obstetrician and a molecular biologist) are in the forefront of their field and have done a great service in helping clinicians become molecular biology literate.

Charles H Rodeck DSc FRCOG
Professor of Obstetrics and Gynaecology
University College Hospital
School of Medicine, London

Preface

The late twentieth century has seen an explosion in our understanding of the structure and function of the genome. Molecular biology, as this has come to be termed, has become a subspecialist scientific discipline, with its own complex concepts and language which may seem to be almost unintelligible to the uninitiated. This new science has found multiple applications to clinical medicine. In obstetrics and gynaecology, in particular, it has revolutionized prenatal diagnosis and is finding applications in other diverse areas such as preimplantation diagnosis, viral DNA detection, investigation of molar pregnancies and paternity testing. Molecular biology techniques are increasingly being used in basic medical research, not just in genetics or the study of genome function but in a wide range of scientific disciplines. The clinical scientist will increasingly find reference to them in the major journals.

The aim of this book is to bring an understanding of the fundamentals of molecular biology, molecular biology techniques and their application to clinical medicine, to clinical obstetricians and gynaecologists. The first section reviews the structure and function of the genome and how alterations to this system may produce human disease. The second section deals with the essential techniques used in molecular biology. These are explained in fundamental detail and assume no prior knowledge of molecular biology on the part of the reader. The third section describes the ways in which these techniques are being applied in the clinical context and how gene therapy might be used, in the future, to treat genetic diseases.

We recommend that the book be read from beginning to end since each section is written on the assumption that the reader has understood what has gone before. Nevertheless, for those who simply must dip into the book, a glossary is provided at the end to help clear the occasional hurdle. We would like to thank Richard Scrivener for help with the photography, Murdoch Elder, Charles Rodeck, Robert Newton, Deborah Henderson, Siobhan Loughna, Simon Forbes and Zehra Ali for their valuable advice and help and Action Research, the Sir Jules Thorn Charitable Trust, the Dunhill Medical Trust, the Association of Spina Bifida and Hydrocephalus and Birthright for the financial support which has enabled us to carry out our research.

Phillip Bennett, Gudrun Moore

Part 1
The Structure and Function of the Genome

1: The history of molecular biology

The discovery of heredity

The fundamental role of the gene in biology was first elucidated by Gregor Mendel. The Austrian monk studied his garden of peas (*Pisum sativum*) in a scientific fashion by crossing different strains. By observing the results he discovered the principles of heredity in the 1860s but his reports were largely ignored until the turn of the century.

In his plant breeding experiments, which occupied much of his life at the monastery, Mendel selected peas with seven pairs of contrasting characteristics, e.g. variations in flower colour or seed shape. In each experiment he crossed plants differing in only one pair of characters. He picked out the hybrids from the first or F1 generation, allowed them to self-pollinate and studied the second or F2 generation. In each of the seven crosses, the F1 generation always resembled only one of the parents. Such characteristics were said to be dominant, e.g. large seeds would be dominant in the progeny of large seeds crossed with small seeds. The characteristic that was hidden was said to be recessive (Fig. 1.1).

The outcome of self-pollination of the F2 generation was even more exciting. He found that a mixture of plants displaying the dominant and recessive characters occurred in a ratio of 3 : 1, with no transitional forms. When the recessive plants in the F2 generation were crossed, all their progeny exhibited the recessive character. However, when the dominant plants were crossed, one-third gave purely dominant progeny with two-thirds giving a similar ratio to that seen in the previous generation (i.e. dominant to recessive of 3 : 1). He explained these results in the following manner. Each plant possesses two 'factors', each of which determines a specific characteristic, but transmits only one factor to its progeny. At any one pollination it is purely a matter of chance which factor is transmitted. This is known as Mendel's first law: 'the law of segregation'. During the formation of the sex cells or gametes, only one of the two factors is found, 'segregated' in the gametes. When two gametes fuse at conception, the factors are doubled, one factor coming from each parent. Each individual possesses two factors for each physical characteristic but these have been independently assorted. If these factors are the same, then for this character or trait the individual is said to be homozygous. If they are different they

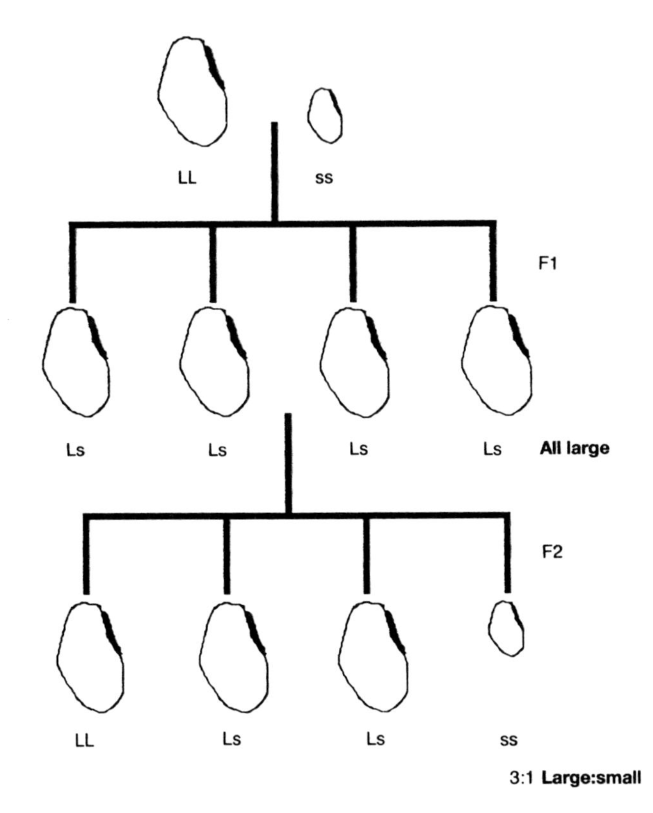

Fig. 1.1 Simple Mendelian genetics demonstrating the inheritance of seed size in the garden pea *Pisum sativum*. The cross between a homozygous large seed (L — dominant) with a homozygous small seed (s — recessive) produces in the first generation (F1) seeds with a heterozygous genotype. Since the large allele is dominant, they all have the large phenotype. In the second generation (F2), reassortment of the alleles results in the genotypes LL, Ls and ss in the ratio 1 : 2 : 1. The phenotypes are both large and small seeds, in the ratio 3 : 1.

are termed heterozygous. In the heterozygous state the character that is displayed is dominant over the recessive, hidden character.

Mendel presented the results from his experiments at the Natural History Society of Brunn in 1865 and they were published in the *Transactions* of the Society a year later. His theories, however, remained largely unknown until 1900 when in the space of 3 months, they were independently rediscovered by three biologists, Hugo de Vries, Professor of Botany at Amsterdam, Carl Correns, a botanist at the University of Tübingen, and Erich von Tschermak-Seysenegg, a research assistant at Esslingen near Vienna. Sadly Mendel died 16 years before the importance of his work became recognized. The term 'gene' was first used to describe Mendel's 'hereditary factors' or characters at the turn of the century by a Danish botanist, Johannsen.

The discovery of chromosomes

The actual physical basis of 'Mendel's theories' was the next area of intense research interest. The building-blocks of living organisms were known to be cells, which contain certain internal structures including, in the case of eukaryote cells, a nucleus. This central organelle was seen to contain thread-like structures, termed chromosomes after chroma for colour, as they would take up certain stains easily. In 1903 Walter S. Sutton and Theodor Boveri independently proposed the 'chromosome theory of heredity'. At cell division these thread-like structures paired up and divided in half, each half ending up in each new cell. This pairing and splitting was consistent with Mendel's 'two-factor' theory.

The discovery of DNA

Deoxyribonucleic acid (DNA) had been identified as a constituent of the nucleus in 1869 by the Swiss scientist Frederick Miescher. It was not until 1946 that DNA was shown to be the material of heredity in bacteria, by Edward Tatum and Joshua Lederberg at Yale University, who first sexually crossed bacteria. They mixed together pairs of different *Escherichia coli* mutants, each bearing different nutritional deficiencies, and obtained progeny bacteria that lacked the nutritional requirements of either parent. The deficiences complemented one another. This showed DNA to be a molecule capable of altering the heredity and, therefore, the genetics of bacteria. It was later shown that the same molecule was the hereditary material of all living cells.

The structure of DNA

In 1953, James Watson and Francis Crick at the Cavendish laboratories in Cambridge postulated the complementary helical structure of DNA. They built a three-dimensional model of DNA based on the energetically most favourable configurations that were compatible with the helical parameters provided by X-ray crystallography data produced by Maurice Wilkins and Rosalind Franklin at King's College in London. In 1962, Crick, Watson and Wilkins were awarded the Nobel prize for this work. Rosalind Franklin died in 1958 and did not receive a Nobel prize as these are not awarded posthumously. Her contribution should not be forgotten. Crick and Watson demonstrated the way in which the three important subunits of the DNA molecule could fit together. They established that DNA is a macromolecular polymer made up of three types of units, deoxyribose (a five-carbon sugar), a phosphate and a nitrogen-containing base. There are two types of bases, purines (adenine and guanine) and pyrimidines (thymine and cytosine). The bases are hydrogen-bonded in specific pairs, adenine (A) to thymine (T) and guanine (G) to cytosine (C). The number of hydrogen bonds dictate that the bases pair up in a fixed fashion with three hydrogen bonds between

G and C and two between A and T. Nucleotides, each composed of a base, sugar and phosphate, polymerize into long polynucleotide chains through 5′ to 3′ phosphodiester bonds.

In 1958, Frank Stahl and Matthew Meselson at the California Institute of Technology demonstrated that DNA replication involved the separation of the complementary strands of the double helix and that identical daughter double helices are generated through the semiconservative replication of DNA. They used density differences to separate the parental DNA molecules from their daughter molecules. They first grew *E. coli* in a medium highly enriched in heavy isotopes, ^{13}C and ^{15}N. Due to its density the heavier DNA, incorporating the isotopes, could be clearly separated from the light DNA by high-speed centrifugation in caesium chloride. When the heavy DNA-containing cells were transferred to a normal 'light' medium and allowed to multiply for one generation, all the heavy DNA was replaced with DNA that weighed half-way between. This indicated a semiconservative replication process with each daughter molecule having one heavy strand from its parent and a new light strand.

The 'Watson–Crick double helix DNA molecule' carries in its biochemical structure the genetic information that allows the exact transmission of genetic information from generation to generation and also simultaneously specifies the amino acid sequence of the polypeptide chains of all proteins needed in the cell. The replication of DNA gives it special properties. Its total information can be copied precisely by separation of the strands followed by synthesis of two completely new strands. This mechanism of DNA replication involves many enzyme systems that include the ability to repair a broken or damaged strand.

Transcription and translation

The work on the enzyme systems of DNA replication was pioneered by Arthur Kornberg at Stanford University. In 1958, DNA polymerase I was isolated and was used to synthesize DNA in a test-tube. Using a DNA strand as a template, DNA polymerase I is able to synthesize a complementary strand by forming phosphodiester bonds between appropriate bases. This work was followed in 1960 by the discovery of RNA (ribonucleic acid) polymerase, an enzyme that makes RNA chains using a template of single-stranded DNA. It was found that eukaryotic cells contain three different RNA polymerases, each with a distinct functional role. Ribosomal RNA (rRNA) is transcribed off ribosomal DNA (rDNA) genes by the enzyme RNA polymerase I. Synthesis of messenger RNA (mRNA) is catalysed by RNA polymerase II, and subsequent addition of poly-adenine tails carried out by the enzyme poly-A synthetase. However, a single form of RNA polymerase is responsible for all RNA synthesis in prokaryotes. At about the same time, messenger RNA was discovered, with the demonstration that it orders the amino acids into proteins.

In 1956, experiments were performed by Francis Crick that confirmed the hypothesis that the genetic messages of DNA are conveyed by its sequence of base pairs. It was found that three adjacent bases (a triplet) coded for each amino acid, constituting a codon. With the use of a synthetic messenger RNA molecule consisting only of uracil (poly-U) the first fragment of the genetic code was found. The amino acid produced from the codon UUU was found to be phenylalanine. By 1966 the complete genetic code was established for all the amino acids and stop codons.

The development of recombinant DNA techniques

In 1961, the genes responsible for conveying antibiotic resistance in bacteria were found to be located on supernumerary chromosomes called plasmids. These later became very important as vehicles or *vectors* to carry foreign and cloned DNA molecules. In 1967, the enzyme DNA ligase was discovered which could join (ligate) DNA chains together. Three years later the first restriction endonuclease was purified. These enzymes, purified from bacteria, cut DNA only at specific sites. In 1978, the Nobel prize for Medicine was awarded jointly to Werner Arber, Hamilton Smith and Daniel Nathans for the discovery of these enzymes and their application to molecular biology.

The knowledge and use of ligase and restriction endonucleases allowed DNA fragments created by a restriction enzyme to be ligated together. The first recombinant DNA molecule was generated at Stanford University by Paul Berg in 1972. One year later foreign DNA fragments were inserted into plasmid DNA to create chimeric plasmids. These plasmids could be functionally inserted into *E. coli* and the foreign DNA replicated and cloned along with the bacterial and plasmid DNA. This breakthrough made it possible to clone any gene or DNA fragment in bacterial cells.

In 1975, E.M. Southern at Edinburgh University developed a technique to blot DNA from agarose gels on to nitrocellulose filters. This became known as a Southern blot. This technique enables restriction-digested DNA to be hybridized to a radiolabelled DNA or RNA probe and has greatly assisted the ability to map DNA molecules and to track and link genes.

In 1977, the first recombinant DNA molecules containing mammalian DNA were produced. In the same year it was discovered that most genes are split, containing introns and exons, and Fred Sanger, Allan Maxam and Walter Gilbert developed techniques to sequence DNA molecules. Walter Gilbert, Fred Sanger and Paul Berg jointly received the Nobel prize for Chemistry in 1980 for their contributions to this field. In 1978, somatostatin, the first recombinant human hormone, was produced by DNA cloning.

In 1981, Judy Chang and Yuet Wai Kan were the first to prenatally diagnose a genetic disorder, sickle-cell anaemia, at the gene level, by restriction enzyme analysis. In 1987, the gene for dystrophin, mutated or deleted in Duchenne muscular dystrophy patients, was cloned and sequenced. In 1989, Lap-Chee Tsui, in Toronto, and Francis Collins, in Michigan, cloned and sequenced the gene whose mutation causes cystic fibrosis. In 1991, a mutation in the amyloid precursor protein gene which causes early-onset familial Alzheimer's disease was found by Alison Goate and John Hardy working in London. These examples are but a few of the breakthroughs that have occurred since the cloning of the first recombinant DNA molecule in 1973. The number of genes localized and cloned is being added to almost every day. Resources are now being made available by scientific research organizations in many countries for the human genome mapping project, which aims to provide a map of the entire human genome and ultimately to sequence all of the genes.

2: The chromosome

Chromosome structure and number

Human cells have their DNA packaged in a nucleus surrounded by a nuclear membrane. The only exception to this is the DNA within mitochondria. Organisms of this type are termed eukaryotes. In contrast, prokaryotes, e.g. bacteria such as *E. coli*, have no nucleus so their DNA is free within the cytoplasm. Eukaryotic cells have a higher level of organization of membranes and organelles, they phagocytose and exhibit a high degree of differentiation and specialization (Figs 2.1 & 2.2).

Chromosomes can be visualized during cell division using specialized staining techniques. At cell division the chromosomes are contracted and can take up stains more easily. Giemsa staining or G-banding has become the most widely used technique for the routine staining of mammalian chromosomes (Fig. 2.3). The chromosome banding patterns obtained are thought to relate to both the structural

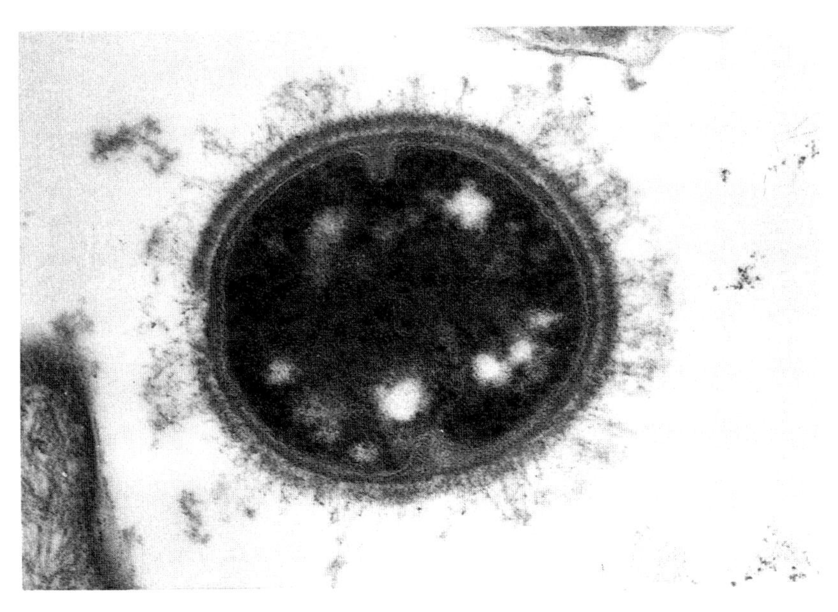

Fig. 2.1 Electron micrograph of a bacterial cell. Prokaryotic cells do not have a nucleus. Photomicrograph courtesy of Dr Tim Ryder, Institute of Obstetrics and Gynaecology, London.

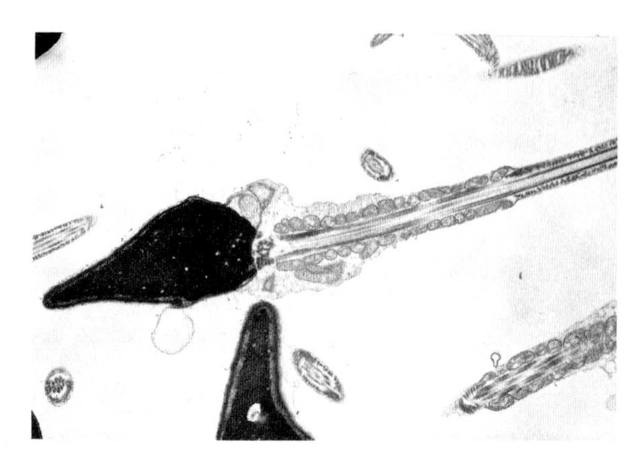

Fig. 2.2 Electron micrograph of a spermatozoon. Genetic material in eukaryotes is contained within a nucleus. In the spermatozoon the nucleus constitutes the majority of the head with only a small amount of cytoplasm. Photomicrograph courtesy of Dr Tim Ryder, Institute of Obstetrics and Gynaecology, London.

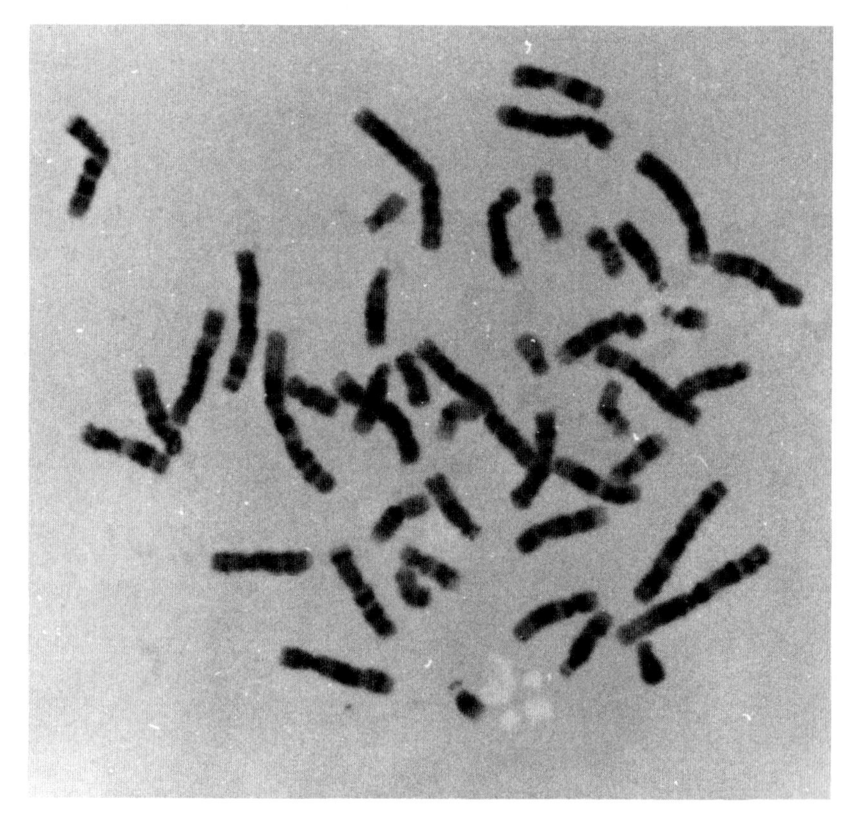

Fig. 2.3 Chromosome spread prepared from an amniocyte at metaphase. The cell was stained with Giemsa to demonstrate the characteristic banding pattern of the chromosomes. Photomicrograph courtesy of Dr Susan Blunt, Institute of Obstetrics and Gynaecology, London.

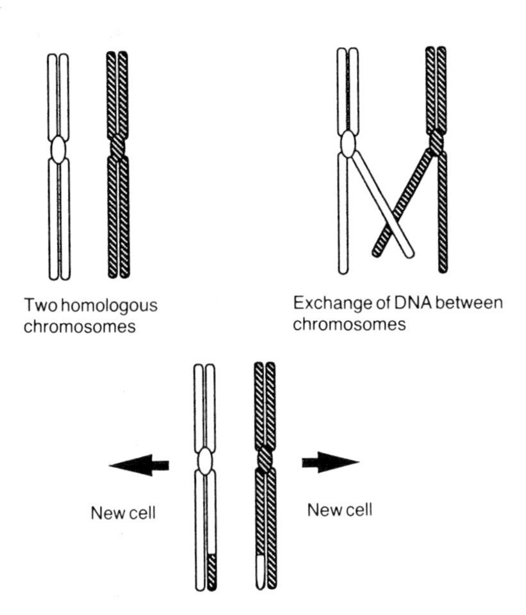

Fig. 2.4 The process of crossing-over between two homologous chromosomes. The net result is exchange of DNA from portions of each chromosome.

and functional composition of the chromosome. The darker bands correlate with pachytene chromomeres, which generally replicate their DNA late in the S-phase. They are found to be rich in the DNA bases, adenine (A) and thymine (T), and appear to contain relatively few active genes. It is thought that they may differ from the light bands in terms of their protein composition. Pachytene is the main stage of chromosomal thickening. The bivalents (paired chromosomes) are in close association and each chromosome is seen to consist of two strands separated at the centromere, termed chromatids; each bivalent is a tetrad of four strands. This is the stage at which crossing-over or chiasma (exchange of homologous material between two of the four strands) occurs (Fig. 2.4).

The chromosomes vary in size. They all contain a pinched portion called a centromere. The centromere is thought to be responsible for the movement of the chromosomes at cell division. Some chromosomes have the centromere in the centre and these are termed metacentric (e.g. human chromosome 1). Others are at the ends, telocentric (no examples in human chromosomes), or very near the end with very short arms, acrocentric (e.g. human chromosome 13). The centromere divides the chromosome into two arms termed the p and q arms. The p arm is usually shorter than the q arm.

Human somatic cells have 46 chromosomes, 22 pairs of autosomes and one pair of sex chromosomes. In males there is an X and a Y chromosome and in females two X chromosomes (Fig. 2.5). The two members of any two autosomes are said to be homologous, one homologue being derived from each parent. Based on their banding patterns

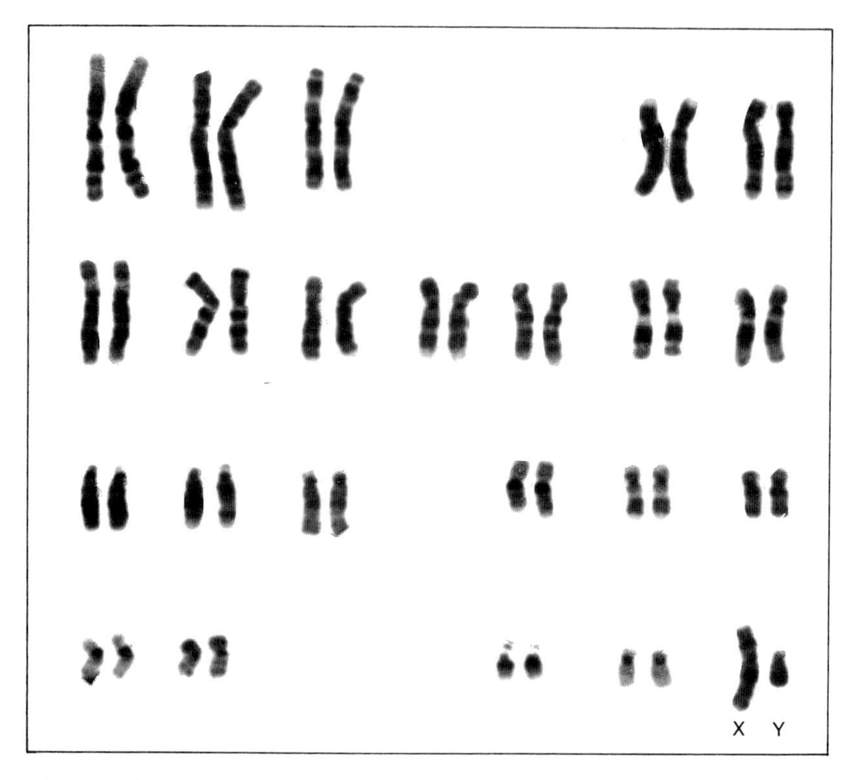

Fig. 2.5 A human male karyotype stained with Giemsa (G-banded). Photomicrograph courtesy of Dr Susan Blunt, Institute of Obstetrics and Gynaecology, London.

each chromosome arm is divided into regions and each region into bands, numbered from the centromere outwards. A given point on the chromosome is designated in the following order: by the chromosome number, the arm symbol, the region number and the band number. Thus the gene for cystic fibrosis is located at 7q31.1 (Fig. 2.6). The number of chromosomes is halved when the gametes are formed. The somatic cells (all those cells other than sex cells or gametes) are diploid, whereas the gametes are haploid. The number of chromosomes per cell varies between species but is constant within species, e.g. the nematode *Caenorhabditis elegans* has 12 chromosomes, the fruit fly, *Drosophila*, has 8, the mouse has 40, the dog has 78 and the carp has 104.

The total extended length of DNA in the chromosomes is approximately 174 cm and this becomes condensed about 10 000 times into the chromosomes that are visible at metaphase. Despite this condensation it still maintains its regional organization and banding. The DNA molecule of the chromosome is complexed with histones (small basic proteins) and non-histone proteins in equal amounts, which together form chromatin. The DNA of a single chromosome is a continuous molecule which forms a single fibre (unineme) of chromatin, even through the centromere. The histones are highly conserved proteins

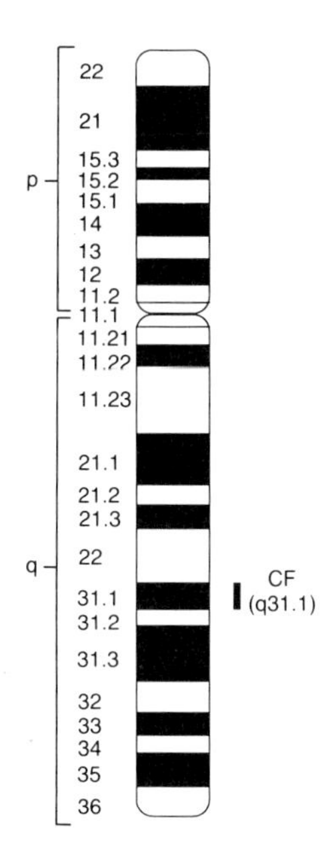

Fig. 2.6 Representation of human chromosome 7 showing the G-banding pattern. The localization of the gene responsible for cystic fibrosis is shown at q31.1.

through a wide range of organisms whereas non-histone proteins are extremely heterogeneous. At metaphase the chromosomes are at their most condensed. Chromosome 1 contracts in length from 15 cm to 10–15 μm and chromosome 21 from 3 cm to 2–3 μm.

Maintenance of chromosome number in somatic cells — mitosis

All cells continue dividing after birth except for neurones. The process is extremely well regulated as errors in nuclear division occur only rarely. The general process of nuclear division in all somatic cells is known as mitosis. During mitosis each chromosome divides into two, leaving the number of chromosomes per cell unchanged. Mitosis can be described in the following stages (Fig. 2.7):

INTERPHASE

When the cell is not actively dividing it is in interphase. Normal cell activity, general metabolic processes and DNA replication occur during this stage. In female cells the Barr body or sex chromatin (the inactive X chromosome) is visible at this time as compacted chromatin. The other chromosomes are metabolically active but not distinguishable from each other. As the cell prepares to divide, the chromosomes condense and begin to separate longitudinally into chromatids, held together at the centromere.

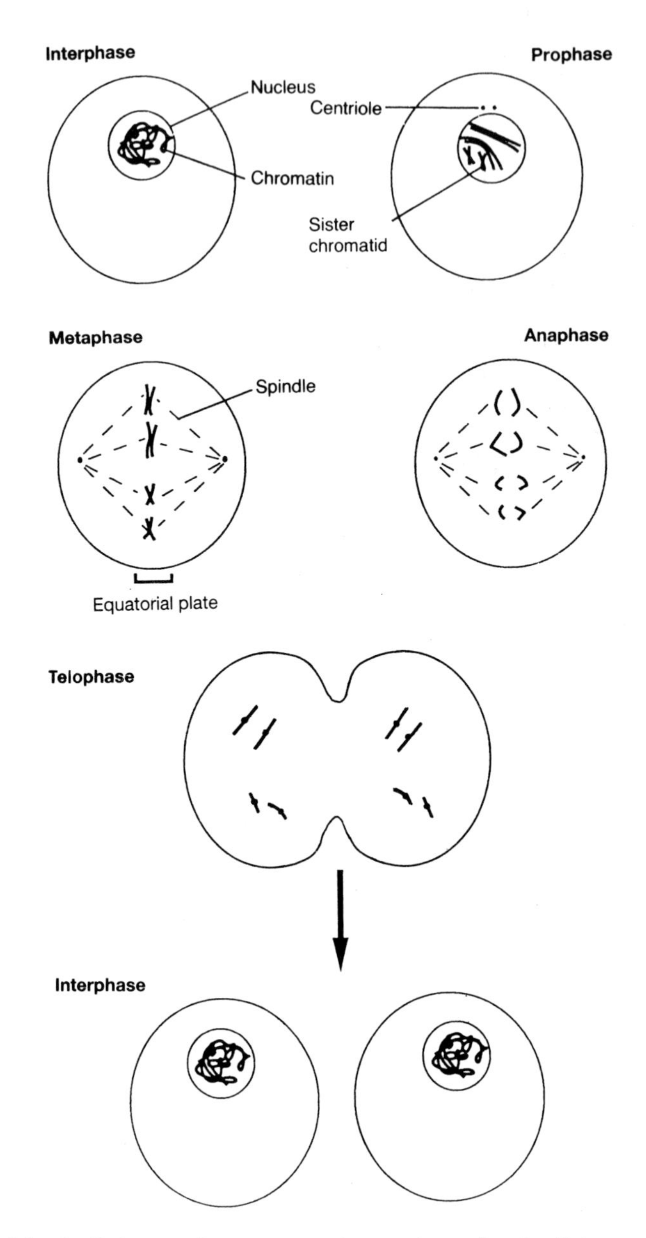

Fig. 2.7 Mitosis. Only two chromosome pairs are shown for simplicity.

PROPHASE

As soon as the chromosomes are discernible the cell has entered prophase. The two sister strands (chromatids) are clearly defined, and the nuclear membrane disappears, dissolving the identity of the nucleus. At the same time the centriole, an organelle found just outside the nucleus, duplicates itself and its two products migrate out to the poles of the cell.

METAPHASE

The condensed chromosomes move to the centre or equatorial plate of the cell. At this stage, metaphase, the individual differences between the chromosomes can be most easily studied as they are at their most separate and condensed. Meanwhile, the spindle has formed. This is a structure that consists of microtubules of protein that radiate from the centriole at the poles of the cell and attach to the centromeres of the chromosomes. These attachment sites are known as kinetochores.

ANAPHASE

At anaphase, the centromere of each new chromosome divides into two daughter chromatids. The two new halves move off into opposite sides of the cell drawn by contraction of the spindles.

TELOPHASE

When the daughter chromatids have moved to the two poles of the cell, telophase has begun. Two new cells are produced from the old single cell, with division of the cytoplasm beginning at the central or equatorial plane. Eventually a complete membrane is formed around the two new daughter cells. Within these new cells, the two groups of daughter chromosomes are surrounded by a nuclear membrane, forming new nuclei. Each of the daughter cells appears as a typical, new interphase cell.

THE MITOTIC CYCLE

Mitotic division takes up only a small part of the life cycle of a somatic cell. Three other stages of the cell cycle are also discerned. After division, the new cell enters a postmitotic stage during which there is no DNA synthesis. This stage is called G1 or Gap 1. The next stage is the S-phase, a period of DNA synthesis in which the DNA content of the cell doubles ready for future cell division. Each strand of DNA present serves as a template for a new exact copy of itself. This is followed by a premitotic period, G2 or Gap 2, which is ended by the onset of mitosis. Studies of cultured human cells show that the complete cycle lasts 12–24 hours, only 1 hour of which is actual mitotic cell division.

SOMATIC RECOMBINATION

Somatic recombination (crossing-over between homologous chromosomes in mitosis) is much rarer than meiotic recombination. Homologous chromosomes do not usually associate except during meiosis. However, there is some evidence that somatic recombination can occur. Individuals that are heterozygous for a disorder, e.g. albinism, can, through recombination at mitotic cell division, show partial mosaicism of albinism, which manifests as patches of white skin.

Production of haploid sex cells — meiosis

Meiosis describes the specialized cell division that occurs when the gametes are formed. Each daughter cell formed by meiosis has a haploid number of chromosomes, with only one representative of a chromosome pair. This is in contrast to mitosis, where the daughter cell is identical in chromosome complement and, therefore, DNA to its parent cell.

There are two successive chromosome divisions. In meiosis I (the reduction division), homologous chromosomes pair during prophase and split apart during anaphase, the centromeres of each chromosome remaining intact. Meiosis I is followed by meiosis II, which is similar to mitosis. The already divided chromosomes are split into chromatids and segregate into daughter cells.

THE FIRST MEIOTIC DIVISION (MEIOSIS I)

Prophase of meiosis I

The prophase of meiosis I is complex and can be considered in several stages (Fig. 2.8):

1 *Leptotene* begins as the chromosomes begin to condense into thin threads. Unlike mitotic chromosomes, they are not smooth in outline but show alternating thick and thin regions. The thicker regions, known as chromomeres, have a specific pattern for each meiotic chromosome.

2 *Zygotene* is the stage of pairing of homologous chromosomes (synapsis). The two partners of each pair are parallel to each other, forming bivalents. This pairing does not occur in mitosis. The X and Y chromosomes are associated only at the tips of their short p arms (termed the pseudoautosomal region).

3 *Pachytene* is the main stage of chromosomal thickening. Each paired chromosome or bivalent consists of two chromatids, displaying a tetrad of four strands. This is the important stage, where crossing-over or recombination occurs (exchange of homologous segments between two of the four strands).

4 *Diplotene* is recognizable by the longitudinal separation that begins to occur between the two components of each bivalent. Although the two chromosomes of each bivalent separate, the centromere of each remains intact, so the two chromatids of each chromosome remain together. During this longitudinal separation the two members of each bivalent are seen to be in contact in several places, called chiasmata (GK. *chiasma* cross). Chiasmata mark the location of cross-overs, where chromatids of homologous chromosomes have exchanged genetic material. In human spermatocytes the average number of cross-overs seen is 50 per cell. Eventually the chromosomes draw apart and the chiasmata draw to the ends of the chromosomes (terminalize).

5 *Diakinesis*, the final stage of prophase, is marked by an even tighter coiling and deeper staining of the chromosomes, as well as terminalization of the chiasmata.

Metaphase, anaphase and telophase of meiosis I

Metaphase I begins as in mitosis with the disappearance of the nuclear membrane and movement of the chromosomes to the centre of the cell. At anaphase I, the two members of each bivalent split, one going to each pole. The bivalents assort themselves independently, so that the chromosomes are sorted into random combinations of paternal and maternal chromosomes, with one representative of each pair going to each pole. The disjunction of paired homologous chromosomes is the physical basis of segregation, whilst the random assortment of paternal and maternal chromosomes in the gametes is the basis of independent assortment. The behaviour of the chromosomes at the first meiotic division provides the physical basis for Mendelian inheritance. By the end of meiosis each product has the haploid chromosome number, which is why meiosis I is often referred to as reduction division.

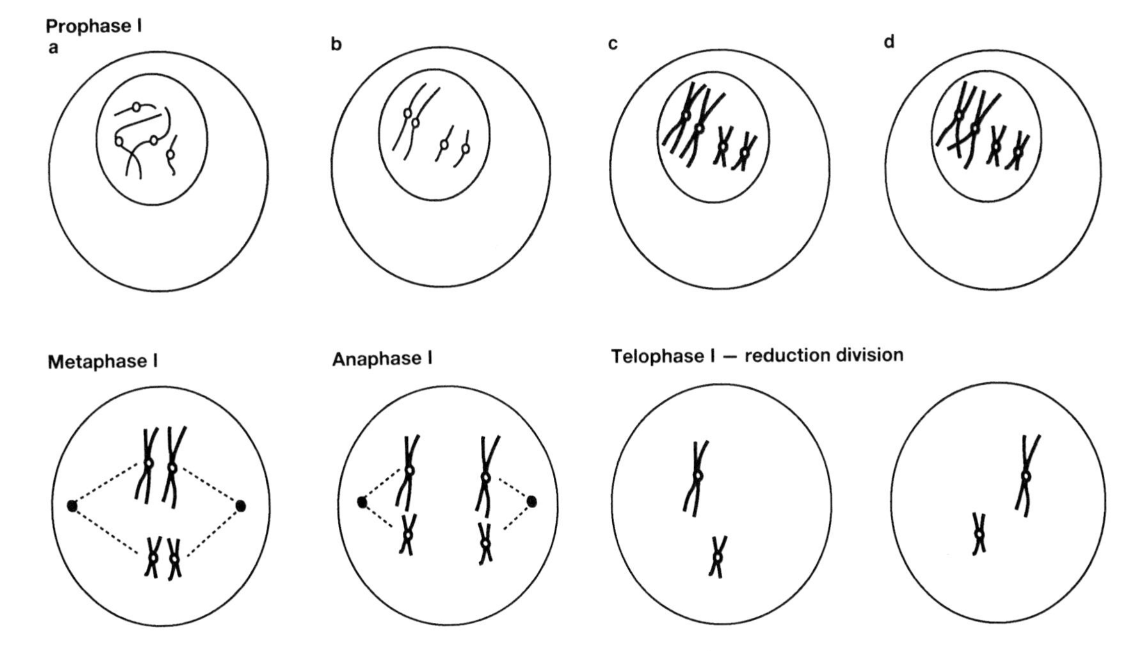

Fig. 2.8 Meiosis I. Only two chromosome pairs are shown for simplicity. (a) Leptotene: chromosomes visible as threads. (b) Zygotene: chromosomes pair. (c) Pachytene: thickening of both strands. (d) Diplotene.

Crossing-over or recombination

A consistent feature of meiosis I is the presence of chiasmata, which hold paired chromosomes together from diplotene through metaphase. These mark the positions of the cross-overs, which are sites where, prior to metaphase, chromatids of homologous segments have exchanged material by breakage and recombination. Although only two chromatids take part in any one event, all four may simultaneously be involved in different cross-overs. Chiasmata may have an important function in holding bivalents together and thus preventing premature disjunction.

Crossing-over causes the genetic material to become reorganized into new combinations, thereby increasing genetic variability. The paternal and maternal chromosomes assort independently at meiosis, allowing for 2^{23} (approximately 8 million) different chromosome combinations in the gametes of a single individual. Crossing-over increases the amount of combinations possible but also increases the likelihood of both favourable and unfavourable new combinations.

THE SECOND MEIOTIC DIVISION (MEIOSIS II)

The second meiotic division follows the first, without a normal interphase or DNA replication (Fig. 2.9). In the oocyte the second meiotic division does not take place until fertilization. In each cell formed by meiosis I, the centromeres now divide and the sister chromatids disjoin, passing to the two poles to form daughter cells. With the exception of areas in which cross-overs have occurred during meiosis I, the daughter cells have identical chromosomes. Thus the end result of the two successive meiotic divisions is the production of four haploid cells, formed by only one doubling of the chromosomal material.

Sexual differentiation

Sexual development depends upon the effect of the sex chromosomes on gonadal differentiation, the correct functioning of the differentiated testis and the response of the end organs to substances produced by the testis. A normal female has two X chromosomes and a normal male one

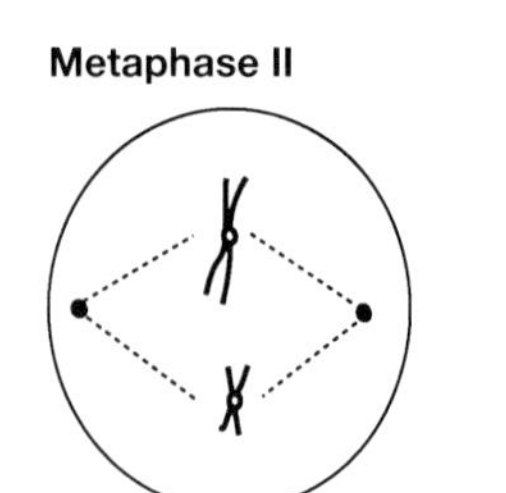

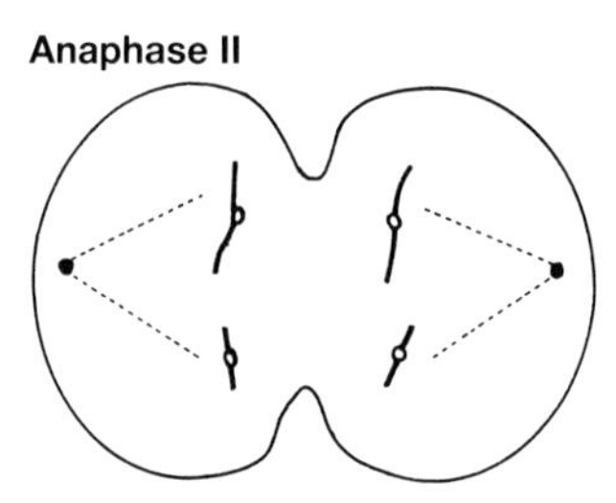

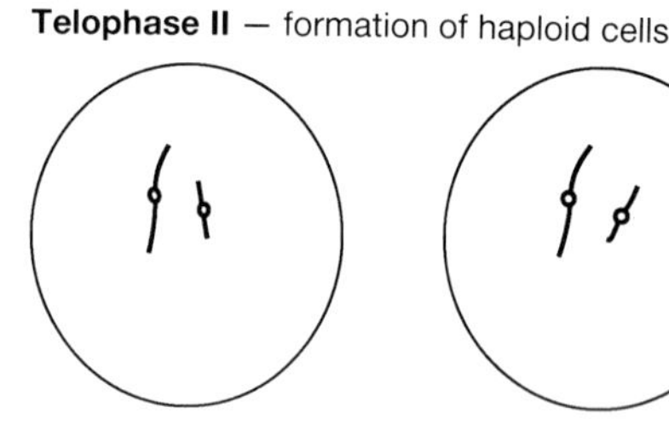

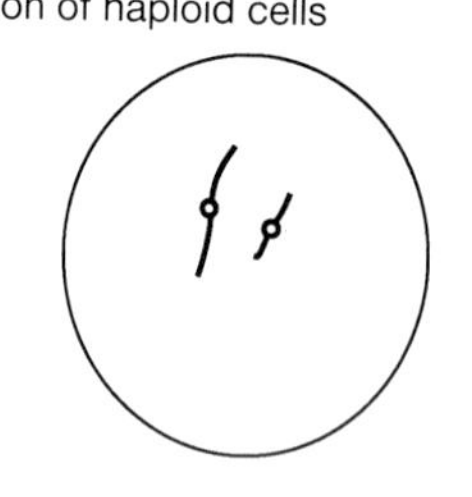

Fig. 2.9 Meiosis II. Only two chromosome pairs are shown for simplicity.

X and one Y. Both the X and the Y chromosomes carry genes other than those for sexual differentiation, although the Y chromosome is smaller than the X. There is a region at the top of the short arm of the Y chromosome which is homologous to the same region of the X. Recombination occurs at meiosis between these two regions in the same way as between autosomes. This is, therefore, often known as the pseudo-autosomal region of X and Y. There is also a homologous region in the mid-part of the long arm of the X chromosome but it is not thought that recombination can occur here. In the normal situation presence of a Y chromosome causes differentiation of the undifferentiated gonads to testes. The Y chromosome carries a gene or genes which function as a testis determining factor (TDF). For many years it was believed that this factor was a male-specific histocompatibility antigen, known as H-Y. This theory was abandoned when it was demonstrated that certain species lack the H-Y antigen. Studies of individuals who were XX, but carried a small translocation from their father's Y chromosome on to X, showed that TDF must be on the long arm of the Y chromosome, just below the X–Y homology region. A gene in this region was found which was highly conserved and which coded for a protein whose structure suggested a DNA-binding transcription regulator. Since this protein contained zinc finger projections it was called ZFY. Again, several individuals were found who lacked ZFY but were undoubtable males. That area of the Y chromosome was searched again and another gene found coding for a protein which is expressed in the developing gonad. A mutation of this gene, 'sex-determining region' of Y (SRY), has been found in mouse which produces XY females. SRY is currently the best candidate for TDF, but this is far from proved.

The testes produce androgens, which cause development of the Wolffian system and a Müllerian inhibitor. The Müllerian inhibitor is a protein. In the human it appears to have a unilateral action, although in cows transplacental transfer of Müllerian inhibitor from a male to a female twin will result in masculinization of the female, producing a freemartin. The Müllerian structures will develop unless inhibited, so, if testes are present in the early embryo and function normally, the embryo will develop as male. If no testes are present, or if they fail to function, the embryo will develop as female. Female development does not require the presence of functioning ovaries. Turner syndrome results from the inheritance of only one X chromosome and no Y chromosome (45X). Although there is no development of either testes or ovaries, the embryo will develop as a female. The appearance of a child with Turner syndrome is essentially female with minor dysmorphic features.

Insensitivity of end organs to normally produced testicular androgens can also result in failure of masculinization and a female phenotype, despite normal testes. Although the Wolffian tissues can respond to testosterone directly, differentiation of the external genitalia requires conversion of testosterone to dihydrotestosterone by the action of 5αreductase. Deficiency in 5α-reductase, occasionally as a result of an autosomal recessive genetic abnormality, will result in partially or scarcely masculinized external genitalia and a female phenotype. Similarly there are abnormalities of testosterone receptor function which may result in failure of masculinization of the embryo.

3: DNA structure and function

The structure of nucleic acids

DNA is composed of a deoxyribose backbone, the three position (3′) of each deoxyribose being linked to the five position (5′) of the next by a phosphodiester bond (Fig. 3.1). At the two position each deoxyribose is linked to one of four nucleic acids: a purine (adenine or guanine) or a pyrimidine (thymine or cytosine) (Fig. 3.2). Each DNA molecule is made up of two such strands in a double helix with the nucleic acid bases on the inside (Fig. 3.3). The bases pair by hydrogen bonding, adenine (A) with thymine (T) and cytosine (C) with guanine (G). DNA is replicated by separation of the two strands and synthesis by DNA polymerases of new complementary strands. With one notable exception,

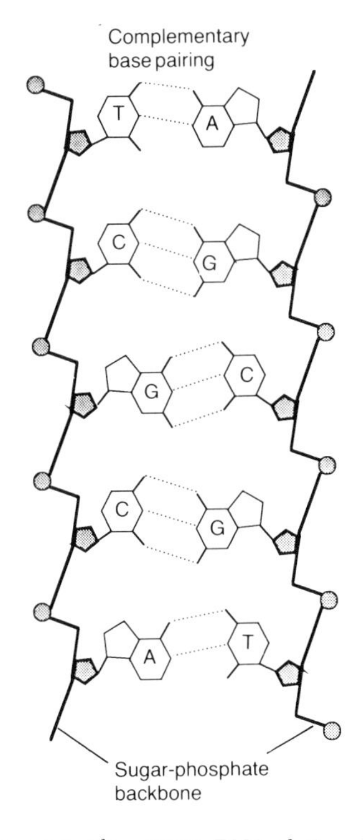

Fig. 3.1 Base pairing of two complementary DNA chains showing the sugar–phosphate backbone.

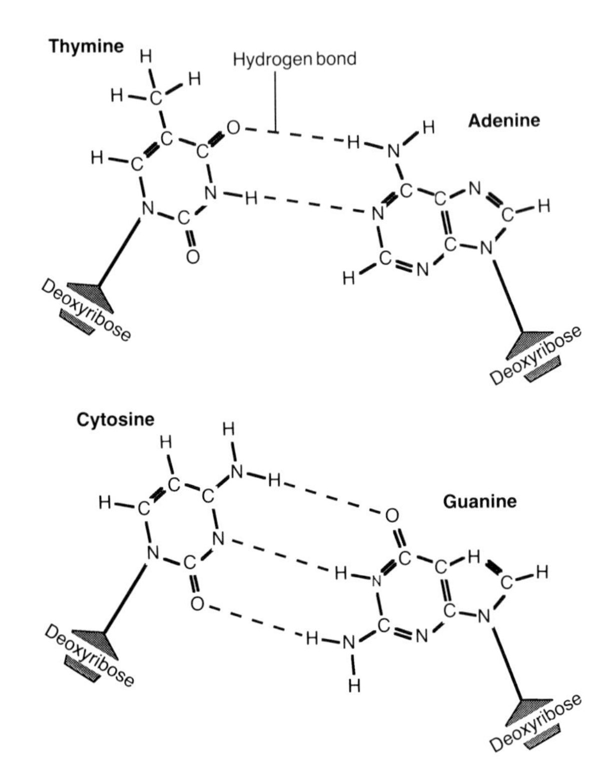

Fig. 3.2 Hydrogen bonding between adenine–thymine and guanine–cytosine base pairs.

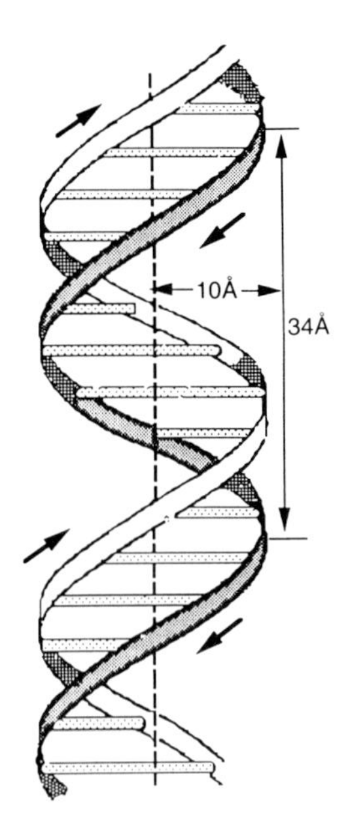

Fig. 3.3 The double helix structure of DNA, as proposed by Watson and Crick in 1953.

the reverse transcriptase produced by viruses, DNA polymerases always add new bases at the 3′ end of the molecule. RNA has a structure similar to that of DNA but is always single-stranded. The sugar moiety consists of ribose, and uracil is found in place of thymine.

Classes of DNA in the genome

The haploid human genome contains approximately 3×10^9 base pairs. It has been estimated that there are 50 000–100 000 genes in the genome, ranging from approximately 1000 to 200 000 bp in size. The structure of much of the remaining non-coding DNA is still unclear but it could be involved in regulation, DNA replication, chromosome pairing and recombination. The human genome can be divided into three classes.

UNIQUE SEQUENCES

These include, but are not exclusively composed of, the protein-coding sequences, present in single copies or only a few copies, comprising approximately 60% of total DNA.

HIGHLY REPETITIVE SEQUENCES

This class covers 10% of the genome or more and is comprised of sequences that are not transcribed and may be repeated hundreds, thousands or even millions of times. The base pairs are clustered in short 4–8 bp tandem repeats. There are extensive heterochromatic highly repetitive regions on chromosomes 1, 9, 16 and the long arm of Y. Slightly larger repeats are found at the centromeres of human chromosomes.

MODERATELY REPETITIVE SEQUENCES

These are present in both dispersed and clustered forms and represent the remaining 30% of the genome. This class contains a number of gene families, functional genes that occur in several to many copies throughout the genome. In humans, the *Alu* family provides one example of a number of dispersed, moderately repetitive sequences, each of which make up 3% or so of the genome. The *Alu* family includes at least 300 000 repeats of a 300 bp sequence. The unit contains the recognition sequence of the bacterial restriction enzyme *AluI*. Other moderately repetitive sequences are substantially longer. *Alu* sequences are also present in other primates and as a shorter 135 bp version in rodents. The function of *Alu* and other moderately repetitive sequences is at present unknown.

Gene structure

Until 1977, the gene was thought of as a segment of DNA containing the code for the amino acid sequence of a polypeptide chain. It is now known that coding sequences, exons, are interrupted at intervals by non-coding sequences, known as introns. These introns are initially transcribed but are not present in mature mRNA and are not translated into protein product. Genes also include extensive flanking regions, important in regulation of the transcription of the gene. An example of the genomic structure, that of the recently characterized cystic fibrosis gene, is shown in Fig. 3.4. The gene responsible for cystic fibrosis has a genomic length spanning 250 kb but a coding length of only 6121 bp spread over 27 exons.

The β-globin gene has been well characterized and is a good example of the general structure of a gene (Fig. 3.5). It consists of three exons divided by two intervening sequences or introns. Upstream (5′) from the gene is an untranslated flanking region containing two specific sequences that play a significant role in regulation: CCAAT, the 'CAT box', usually found about 70 bp upstream from the start of transcription, and TATA, the 'TATA box', found about 30 bp upstream from the start of transcription. 'TATA' and 'CAT' boxes are believed to function in defining the exact nucleotides from which initiation of transcription occurs. The 'TATA box' helps direct RNA polymerase II to the correct site for initiation of transcription. The precise function of the 'CAT box' is unknown. Downstream (3′) is a second flanking region, containing a sequence, AATAAA, that appears to be a signal for the addition of the poly-A tail to the end of the mRNA strand, and this is transcribed but not translated. It is not, however, a transcription-termination signal. It most probably serves as a signal for a nuclease to clip the RNA chain at a specific site 10–15 bases further down. A second enzyme, poly-A

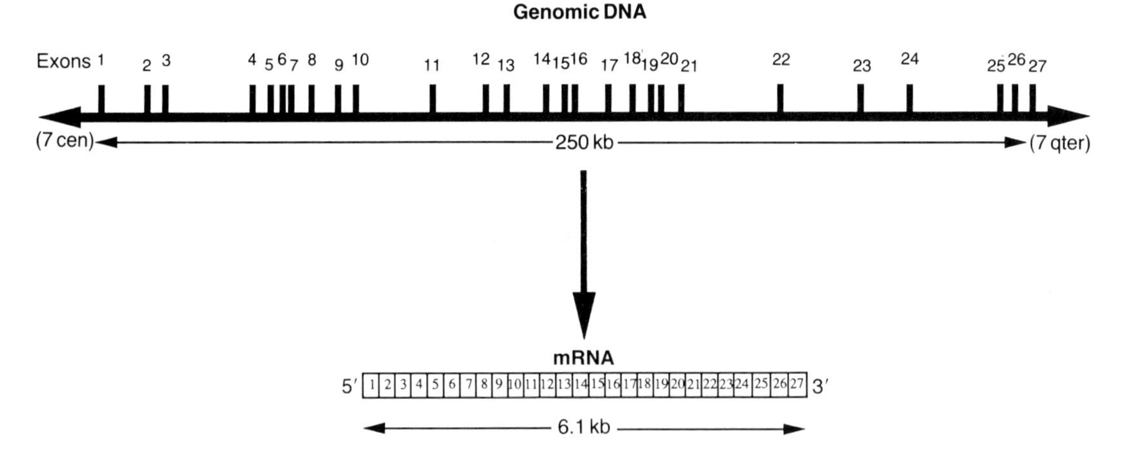

Fig. 3.4 Organization of the cystic fibrosis gene exons. The 27 exons that constitute a 6.1 kb messenger RNA are spread over 250 kb on the chromosome.

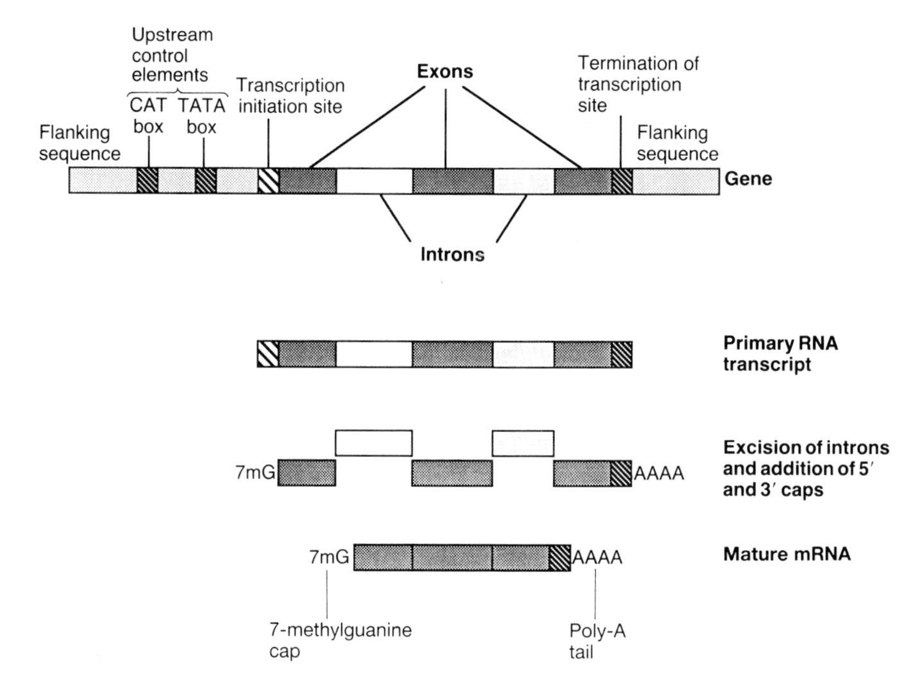

Fig. 3.5 Structure of a generalized human gene, showing the results of transcription and processing to form the mature mRNA.

synthetase, then adds the variable-length poly-A tail to the mRNA.

Typical genes of vertebrates have coding sequences split by intervening sequences. An exception to this is the antiviral agents, interferons, a family of genes on chromosome 9. In lower eukaryotes, splitting of coding sequences with introns is less common and it is not seen in prokaryotes. Intron–exon patterns tend to be conserved during evolution, e.g. the α- and β-globin genes are believed to have arisen as a duplication of a primitive precursor 500 million years ago, with the two introns in the same location as they are today. Within the exons, alterations in sequence occur slowly by the fixation of rare advantageous mutations through natural selection. The average rate of substitutions per codon per year is one in 10^9 bp. Mutations in intron sequences, except at splice junctions, rarely confer any advantage or disadvantage on the individual. There is no selective pressure against minor variations in intron sequence. Such minor variations are, therefore, more common in intron than in exon sequences.

Gene families

A number of 'families' of closely related genes have been defined. Two families (or a single extended family), the α-like and the β-like globin gene clusters, have been described on chromosomes 16 and 11 respec-

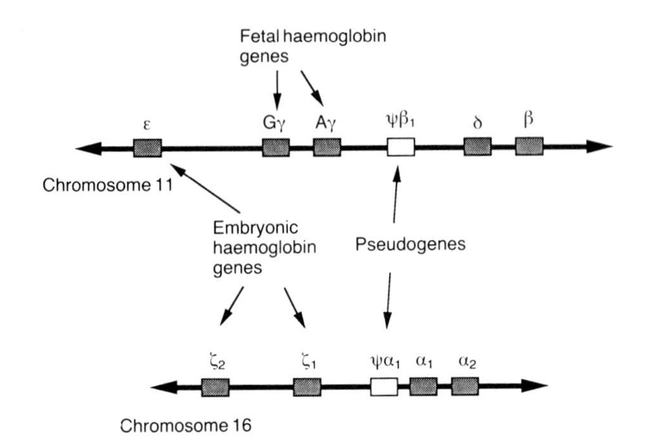

Fig. 3.6 The α- and β-globin gene families. Genes in these families are clustered together on their respective chromosomes and are apparently ordered in the sequence in which they are activated during development.

tively. In each of these two linked clusters, the genes that code for these closely related proteins are expressed at different developmental stages. The sequence of the genes along the chromosome is the same as the order of their activation during development, with spacer sequences separating them (Fig. 3.6). Albumin and α-fetoprotein are two genes which, although coding for two different proteins, are clustered and show homology. Some gene families, including both globin clusters, contain pseudogenes. Pseudogenes show striking homology to functional genes but for some reason, possibly since they display evolutionary mutations, are not functional.

Transcription — the synthesis of messenger RNA

The central dogma of molecular genetics is that a gene, consisting of double-stranded DNA, is transcribed into single-stranded messenger RNA (mRNA), and this is translated into a protein. In general, the concept of one gene one protein holds true, although there are examples of proteins coded for in parts, by genes on different chromosomes, and there are genes which code for RNA only, e.g. the transfer RNA (tRNA) genes. The amino acid sequence of the protein is coded in the DNA by triplets of nucleic acid bases. Each three bases, known as a codon, codes for a single amino acid (Fig. 3.7). In most cases a single amino acid may be coded for by a variety of codons. The genetic code is, therefore, said to be degenerate. There are also three codons which cause termination of transcription — nonsense or stop codons.

During transcription, the RNA polymerase which constructs the mRNA reads from the DNA strand complementary to the RNA molecule. This is known as the antisense strand whilst the opposite strand, which has the same base pair composition as the RNA molecule (except

First position	U	C	A	G	Third position
U	Phe	Ser	Tyr	Cys	U
	Phe	Ser	Tyr	Cys	C
	Leu	Ser	STOP	STOP	A
	Leu	Ser	STOP	Trp	G
C	Leu	Pro	His	Arg	U
	Leu	Pro	His	Arg	C
	Leu	Pro	Gln	Arg	A
	Leu	Pro	Gln	Arg	G
A	Ile	Thr	Asn	Ser	U
	Ile	Thr	Asn	Ser	C
	Ile	Thr	Lys	Arg	A
	Met	Thr	Lys	Arg	G
G	Val	Ala	Asp	Gly	U
	Val	Ala	Asp	Gly	C
	Val	Ala	Glu	Gly	A
	Val	Ala	Glu	Gly	G

(Second position: U, C, A, G)

Fig. 3.7 The genetic code.

that DNA has T in place of U), is the sense strand. Gene sequences are expressed as the sequence of the sense strand of DNA, although it is in fact the antisense strand which is read.

After transcription of a gene, intron regions are excised from the premessenger RNA and the coding regions are ligated together to make a complete, mature messenger RNA molecule. This process is termed RNA splicing; the splice junctions of introns are characterized by 'consensus sequences' of bases, which invariably include GT at the 5′ end and AG at the 3′ end. Splicing can be disrupted by mutations that change one of these sequences or alter a sequence elsewhere so that it closely resembles a splice junction. The mature RNA molecule has an additional chemical modification, known as capping, at the 5′ end to protect against ribonucleases (RNases). A polyadenine tail is added to most mRNA molecules at their 3′ end.

Translation — the synthesis of proteins

Once in the cytoplasm the mRNA message is translated into protein by a ribosome (Fig. 3.8). Ribosomes, consisting of a complex bundle of proteins and ribosomal RNA, attach to mRNA at the 5′ end. Protein synthesis begins at the amino terminal and amino acids are sequentially added at the carboxyl end. Amino acids are brought into the reaction by specific tRNA molecules. Each tRNA is a single-stranded molecule which folds in a way that allows complementary base pairing between parts of the same strand. The specific configuration allows the tRNA molecule to bind to its specific amino acid. At one end of the molecule three bases remain which are unpaired and complementary to the

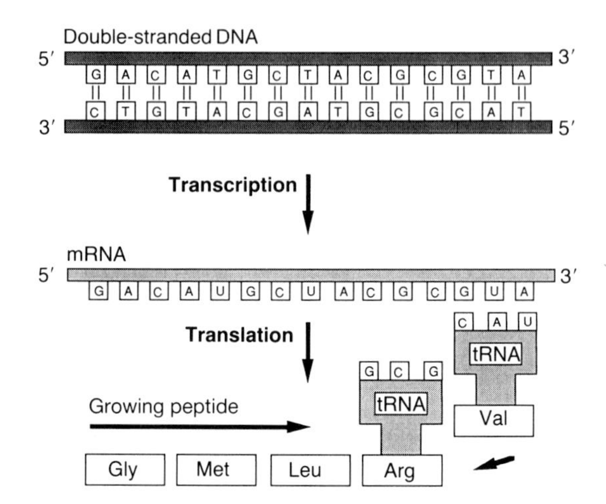

Fig. 3.8 Messenger RNA is transcribed from the antisense strand of genomic DNA. The introns are spliced out and the mature messenger RNA migrates into the cytoplasm. At the ribosome the messenger RNA is translated into protein. Addition of successive amino acids is mediated by transfer RNA molecules.

codon coding for the amino acid. This anticodon binds to the codon of the mRNA and places the amino acid in the correct sequence of the protein. Usually several ribosomes translate a single mRNA molecule at any one time. Coding regions of a gene can be on either strand of the DNA molecule. Since three bases constitute a codon, transcription could begin in any one of three phases on each strand. The actual phase which is translated into protein is known as the reading-frame.

DNA mutations

Any change in a sequence of genomic DNA constitutes a mutation. Normally DNA replication is accurate but any errors that do occur in the germ cells may be copied through generations. Phenotypic effects of a mutation depend on the nature of the alteration, if any, in the corresponding gene pool. There are three main mechanisms of mutation: substitution, deletion and insertion.

SINGLE-BASE SUBSTITUTIONS

Substitution is a single-base change (Fig. 3.9). If the substitution changes a codon, the wrong amino acid will be incorporated into the growing protein chain. Even a single amino acid change may make a radical difference to the folding of a protein and destroy its function. Since the genetic code is degenerate not all substitutions lead to changes in the amino acid sequence. In many cases a change in the third base of a codon will not change the amino acid being coded for. The best known example of a single-base substitution is the sickle mutation. In this case

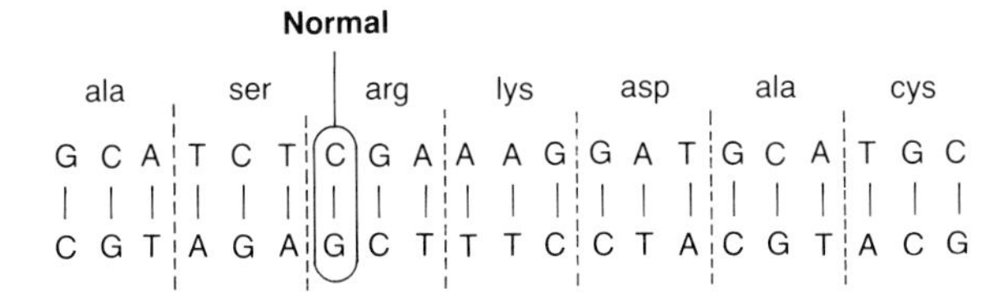

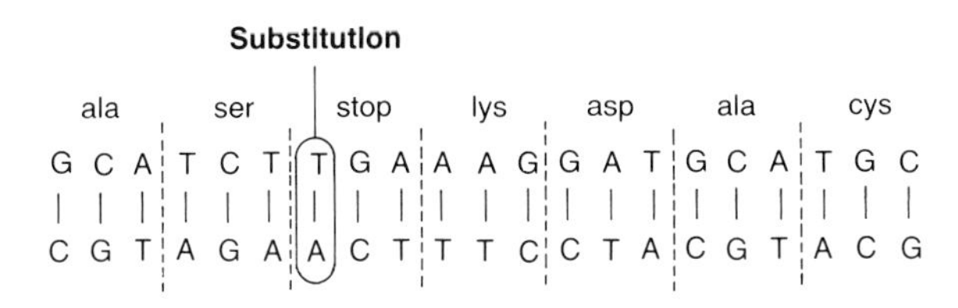

Fig. 3.9 Substitution. One or more bases are replaced. This does not alter the length of the gene but may alter codons, leading either to incorrect amino acids being incorporated into the protein or to premature termination of transcription if a stop codon is introduced. Due to redundancy in the genetic code, some substitutions do not alter the encoded amino acid and are therefore silent mutations.

the β-globin gene carries a single substitution of T for A, which changes a codon from GAG, coding for valine, to GTG, coding for glutamine. The single amino acid substitution changes the stereochemistry of β-globin to the form which sickles in conditions of low oxygen tension. In this case a single-base change also deletes a recognition site for the restriction enzyme *MstII*. If genomic DNA is digested with this enzyme and a probe specific for the β-globin gene used on a Southern blot, it will detect a band larger than usual if the sickle mutation is present. This can be used for DNA diagnosis of sickle disease (see Chapter 14). In most cases of single-base mutations, however, there is no change in the restriction map around the gene and more sophisticated techniques are needed to identify them.

DELETIONS AND INSERTIONS

Deletions and insertions may be a single base, several bases, an entire exon or gene or several genes (Figs 3.10 & 3.11). If the deletion or insertion is in multiples of three base pairs and coincides with the correct reading-frame, the protein may be synthesized with only minor alterations. However, if the deletion or insertion is out of frame, the remainder of the protein may be synthesized with a different sequence of amino acids. This is termed a 'frame-shift' mutation. This frame-shift mutation will code for a nonsense amino acid sequence until one of the stop codons is encountered. Transcription of a protein ceases after the site of a stop codon in the RNA template. A stop codon can be introduced either as a result of a point mutation or by a frame-shift mutation bringing a termination codon into frame. This will cause premature

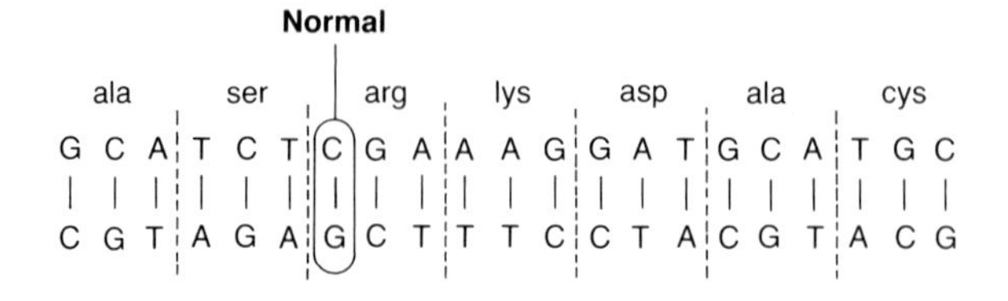

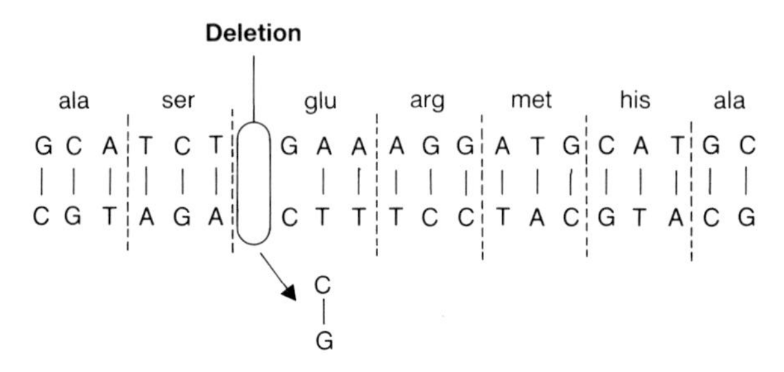

Fig. 3.10 Deletion. Deletion of a multiple of three bases will cause loss of amino acids without altering the remainder of the protein sequence. Other deletions will disrupt the reading-frame, leading to either abnormal amino acid incorporation or termination of protein synthesis. Deletions may vary in size from single base pairs to whole regions of chromosomes.

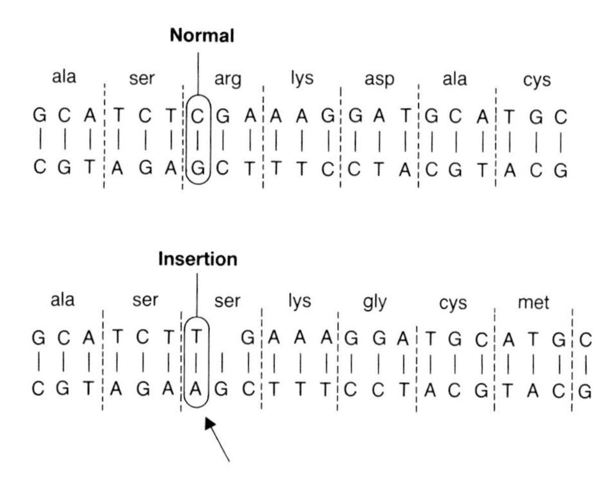

Fig. 3.11 Insertions. Insertions of a multiple of three bases will cause incorporation of extra amino acids without altering the remainder of the protein sequence. Other insertions will disrupt the reading-frame, leading to either abnormal amino acid incorporation or termination of protein synthesis.

cessation of transcription and result in a truncated protein product. This is termed a chain-termination mutation.

Deletion of a region of the gene which does not affect the reading-frame, a deletion which is a multiple of three base pairs, may produce a protein which retains some of its original function. The X-linked disorder Duchenne muscular dystrophy is caused as a result of various deletions of the coding region of the gene encoding the structural protein, dystrophin. The α-thalassaemias are caused by deletions in the α-globin gene cluster. In some cases deletions causing thalassaemia may be large enough to involve several genes. There is a variant of β-thalassaemia caused by deletion of a 600 bp region of the β-globin gene. This deletion shortens the distance between two *BglII* recognition sites (see Chapter 14). On Southern blotting the β-globin gene probe detects a shorter fragment if this β-globin mutation is present. Unfortunately, there are a large number of other mutations which cause β-thalassaemia which do not alter a restriction fragment and cannot be identified so simply.

OTHER MUTATIONS

Mutations in exon–intron boundary sequences may affect the normal mechanism by which introns are excised and exons spliced together. This will usually result in a complete failure of gene product synthesis. Changes in CAT and TATA boxes or other regulatory sequences, however, are more likely to result in reduced transcription.

4: Human genetics

Chromosomal abnormalities

Chromosomes, whether autosomes or sex chromosomes, are subject to two kinds of change, numerical and structural. Numerical chromosomal abnormalities usually arise as a result of non-disjunction. This is an error in cell division in which there is failure of the paired chromosomes or sister chromatids to separate or disjoin at anaphase. This error may occur at a mitotic division or during the first or second meiotic division. Changes in chromosome number are termed aneuploidy. An increase in the chromosome complement which is a multiple of the haploid number is termed polyploidy.

ANEUPLOIDY

An aneuploid is an individual that has a chromosome number that is not an exact multiple of the haploid number of 23, e.g. 45 (monosomy) or 47 (trisomy). During meiosis one of a pair of homologous chromosomes fails to separate (non-disjunction), resulting in an unequal distribution of one pair of homologous chromosomes to the gametes. One haploid cell has two copies of one chromosome whilst the other lacks a chromosome.

Monosomy

Autosomal monosomy is usually incompatible with embryonic development. Most monosomic pregnancies abort in the first trimester. Surviving individuals with monosomy of an autosome are rare. About 97% of embryos lacking a sex chromosome also abort but a few survive and show the characteristics of Turner syndrome: short stature, webbed neck, absence of sexual maturation and a broad chest with widely spaced nipples. The incidence of 45X Turner syndrome in the newborn population is approximately one in 10 000. The Turner sex chromosome abnormality is the most common cytogenetic abnormality seen in fetuses that are spontaneously aborted and accounts for 18% of abortions that are caused by chromosomal error. The error of non-disjunction is usually in the paternal gamete and the incidence is not related to maternal age.

Trisomy

If three chromosomes are present instead of the normal pair, the cell or fetus is trisomic (Fig. 4.1). The error is once again in the gamete, with the sex cell resulting with 24 instead of 23 chromosomes. The error is usually in the maternal gametes. Autosomal trisomy generally causes abortion of the pregnancy in the first trimester. Of autosomal chromosomes only trisomy of 21, 18 and 13 are compatible with fetal development to term. Each is associated with developmental abnormalities and mental retardation. Trisomy 21 is Down syndrome, trisomy 18 Edwards syndrome and trisomy 13 Patau syndrome. Down syndrome is the most common and its incidence increases with maternal age from one in 1000 for a mother aged 25 to one in 100 for a mother aged 40. Down syndrome individuals have a typical facial appearance, including a flat broad face, oblique palpebral fissures, epicanthus, speckling of the iris and a furrowed lower lip. Individuals with Down syndrome may survive to middle age. Edwards syndrome and Patau syndrome are associated

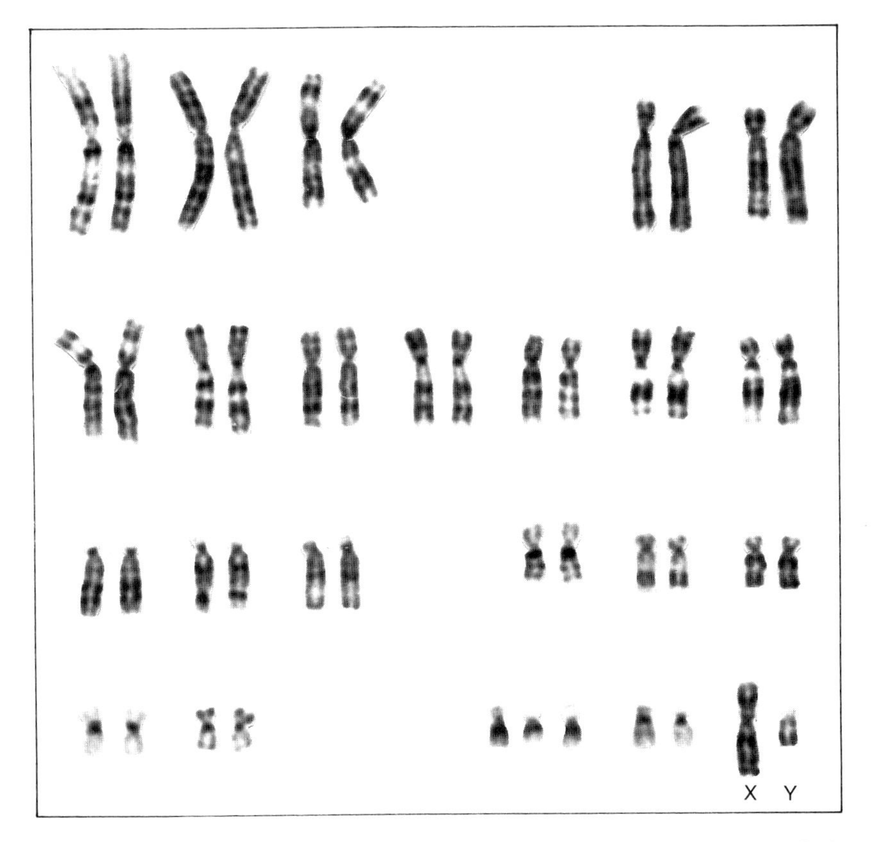

Fig. 4.1 A human male karyotype stained with Giemsa (G-banded). This individual has an additional chromosome 21 which gives the Down syndrome phenotype. Photomicrograph courtesy of Dr Susan Blunt, Institute of Obstetrics and Gynaecology, London.

with more severe dysmorphology. Affected neonates rarely live for more than a few days.

The fact that these three trisomies are compatible with intrauterine development and live birth could be reflective of the type or amount of coding genes on chromosomes 13, 18 and 21. It is known that chromosomes 13 and 18 apparently contain fewer coding genes than other human chromosomes. Over-expression of the genes present on these chromosomes as a consequence of a third copy must relate in some way to the malformations seen in the fetus.

Trisomy of the sex chromosomes is a more common condition and, as there are no characteristic physical findings in infants and children, may not be diagnosed until adolescence. Females with 47XXX have a normal appearance and are usually fertile but may be mentally retarded. Males with 47XXY (Klinefelter syndrome) have small testes, hyalinization of seminiferous tubules and aspermatogenesis, and are often tall with disproportionate lower limbs. Males with 47XYY are also tall and may have a problem with impulse control but are normal in appearance. All three are found at an incidence of 1 in 1000.

Tetrasomy and pentasomy are rare but four or five sex chromosomes have been found in some mentally retarded individuals. Multiple sex chromosome complexes have also been reported: in females XXXX and XXXXX; in males, XXXY, XXYY, XXXYY and XXXXY. Usually the greater the number of X chromosomes present, the greater the severity of the mental retardation and physical impairment. The extra sex chromosomes do not accentuate male or female characteristics.

Mosaicism

Individuals with one or more cell lines with a different karyotype are said to be mosaics. Either the autosomes or the sex chromosomes may be mosaic. Usually individuals who have a mosaic of normal and aneuploid cells have less serious malformations than in persons with an aneuploidy. For example, the features of Turner syndrome are not so evident in a 45X/46XX mosaic female as in the usual 45X females. Mosaicism usually arises from non-disjunction during early mitotic cleavages.

POLYPLOIDY

Polyploid cells contain multiples of the haploid number of chromosomes, e.g. 69 or 92 chromosomes. Polyploidy is a significant cause of spontaneous abortion. The most common type is triploidy. This can result from the second polar body failing to separate from the oocyte or from an ovum being fertilized by two sperm (dispermy). Some of these fetuses are born but all die within a few days. Tetraploidy probably occurs during preparation for the first cleavage division. Normally, each chromosome replicates and then divides into two. Consequently, when the zygote divides each blastomere contains 46 chromosomes. If the chromosomes divide but the zygote does not undergo cleavage at this stage, the zygote will contain 92 chromosomes.

STRUCTURAL CHROMOSOME ABNORMALITIES

Most structural abnormalities result from chromosome breaks induced by various environmental factors, e.g. radiation, drugs and viruses. The type of abnormality that results depends upon what happens to the broken pieces. Translocation is the transfer of a piece of one chromosome to a non-homologous chromosome (Fig. 4.2a). If two non-homologous chromosomes exchange pieces, the translocation is reciprocal. A translocation which does not result in the loss of genetic material is termed balanced and does not necessarily lead to abnormal development. Individuals with balanced translocations reported between chromosomes 21 and 14 have been found to be phenotypically normal. Such persons are termed translocation carriers. If, during meiosis, normal chromosome 21 segregates with the chromosome 14 carrying extra material, that gamete will carry two copies of some chromosome 21 material. After fertilization and combination with another normal gamete, the embryo will have three copies of that chromosome 21 material. This may result in the Down syndrome phenotype. Approximately 4% of Down syndrome is due to translocation. The risk of this type of Down syndrome is independent of maternal age in those that carry the translocation.

When a chromosome breaks, a portion of the chromosome may be lost, resulting in a deletion (Fig. 4.2b). A partial terminal deletion from the short-arm end of chromosome 5 causes *cri-du-chat* syndrome.

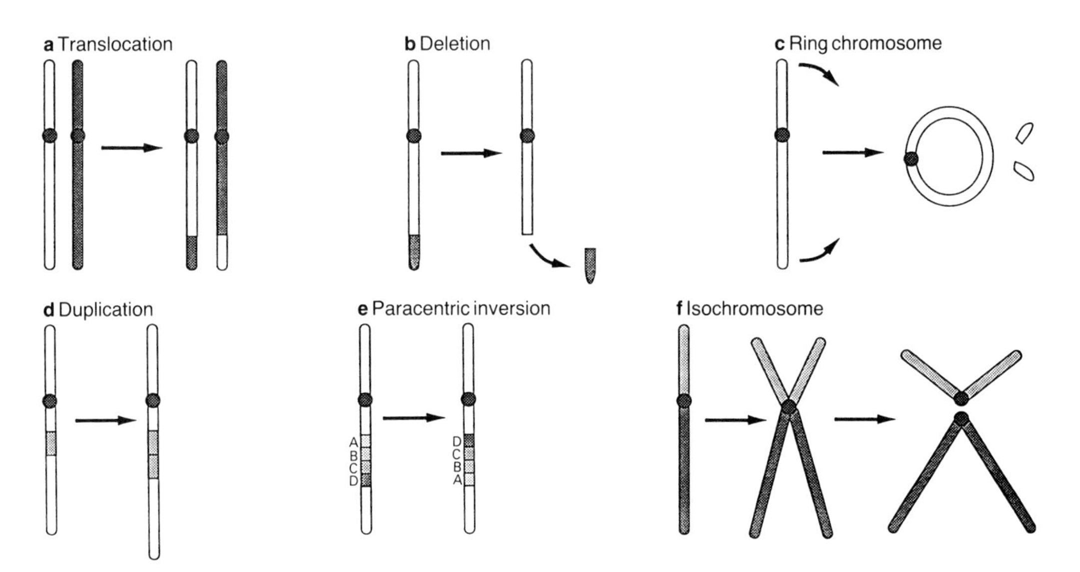

Fig. 4.2 Chromosome rearrangement: (a) translocation; (b) deletion; (c) ring chromosome; (d) duplication; (e) paracentric inversion; (f) isochromosome.

Affected individuals have a weak cat-like cry, microcephaly, severe mental retardation and congenital heart disease. A ring chromosome is a type of deletion chromosome from which both ends have been lost and in which the broken ends have rejoined to form a ring-shaped chromosome (Fig. 4.2c).

Duplication of a part of the chromosome can also occur and can be either within a chromosome, attached to a chromosome or as a separate fragment (Fig. 4.2d). Duplications are more common than deletions, and because there is no loss of genetic material they are less frequently associated with physical or mental abnormalities. Duplications may involve non-coding DNA only, a single gene or a series of genes.

Reversal of a segment of a chromosome is termed an inversion (Fig. 4.2e). Paracentric inversion is confined to a single arm whereas pericentric inversion involves both arms. Pericentric inversion has been described in connection with Down syndrome and other abnormalities. When the centromere divides transversely instead of longitudinally, the abnormality that results is termed an isochromosome, which appears as the most common structural abnormality of the X chromosome (Fig. 4.2f). Patients with this abnormality are often short in stature and have many features in common with Turner syndrome. These characteristics are related to the loss of the short arm of one of the X chromosomes.

Single-gene defects

Single-gene disorders result from mutations in specific genes. If the mutation is phenotypically expressed, the mutation will be on both homologous chromosomes in recessive conditions, or on just one in dominant conditions. Single-gene defects exhibit simple patterns of Mendelian inheritance in the families in which they are inherited. They are described in three groups: autosomal dominant, autosomal recessive and X-linked (which may be dominant or recessive).

AUTOSOMAL DOMINANT INHERITANCE

Autosomal dominant inheritance is the easiest to follow in a pedigree (Fig. 4.3). As in all autosomal disorders, the transmission of the mutation is independent of the sex of the individual. All homozygous and heterozygous individuals are affected. When two heterozygous parents have children, each child has a 75% risk of being affected. When one parent only is affected, then the risk is reduced to 50%. However, a phenotypically unaffected family member cannot transmit the defect. One of the features of autosomal dominant inheritance is variable expression. A mildly affected parent can have a severely affected child. Clearly an autosomal dominant trait which has a severe effect upon survival or fertility would not persist in the population without the influence of variable expression.

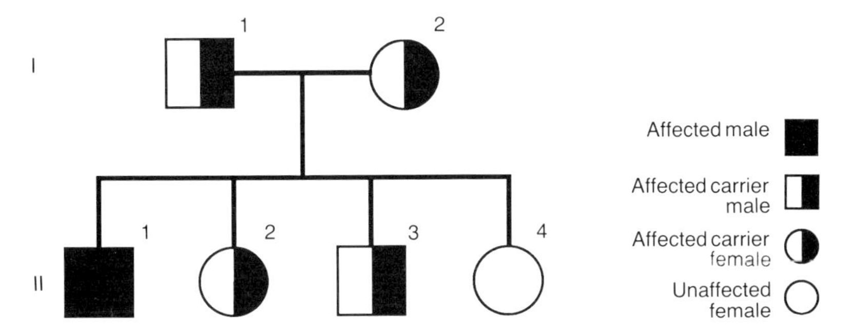

Fig. 4.3 Pedigree of a family displaying inheritance of an autosomal dominant condition. Both homozygous and heterozygous inheritance of the mutation causes an affected phenotype.

Negative selection pressure is not a factor of autosomal dominant diseases which have an onset later in life. Huntington disease is a fatal disorder causing degeneration of the nervous system. It is a late-age onset disorder, 50% of affected individuals displaying the disease at 40 years of age. This mutated gene is localized to the top of the p arm of chromosome 4 and was one of the first whose chromosome position was determined using DNA probe markers. This linkage has enabled DNA diagnosis of potentially affected individuals at an early age.

AUTOSOMAL RECESSIVE INHERITANCE

An autosomal recessive trait is phenotypically expressed only in homozygotes who have received the mutation from both parents (Fig. 4.4). If both parents are carriers, each child has a 25% chance of being affected and a 50% chance of being a carrier. The best-known examples of autosomal recessive diseases are the thalassaemias (discussed in greater detail in Chapter 14). There are a large number of different mutations which may cause thalassaemia. It is, of course, possible for an affected

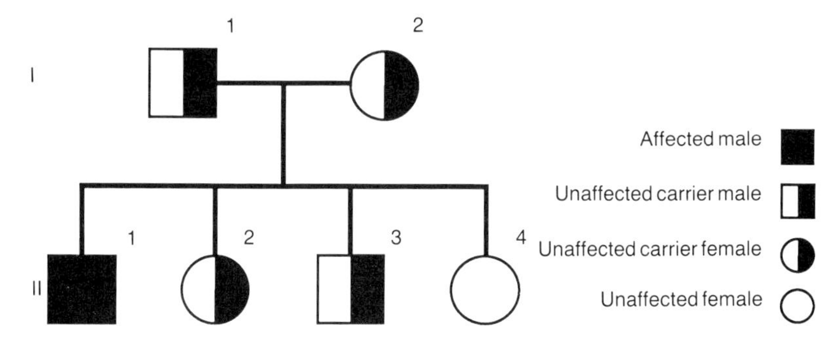

Fig. 4.4 Pedigree of a family displaying inheritance of an autosomal recessive condition. Only individuals who are homozygous for the mutation have the affected phenotype.

individual to inherit a different mutation in the same gene from each parent. Although certain mutations tend to predominate in certain populations, the increase in racial cross-breeding in the late twentieth century has served to complicate the situation.

The most frequent autosomal recessive disorder in Caucasians is cystic fibrosis (CF). This is a childhood condition displaying malfunctions in several exocrine secretions, including pancreatic and duodenal enzymes, sweat chlorides and bronchial secretions. The incidence of the disease is 1 in 2000. One in 20–25 individuals is a carrier. The gene for cystic fibrosis was localized in 1985 to the q arm of chromosome 7, and in 1989 the gene was cloned. Until the gene was cloned in 1989 it was believed that there was only one cystic fibrosis mutation in the Western European population. It is now known that a single codon deletion (coding for phenylalanine) is the mutation in 70% of CF genes with a variety of other mutations in the remaining 30%.

X-LINKED INHERITANCE

The inheritance of recessive genes on the X chromosome follows a clear pattern (Fig. 4.5). With all X-linked diseases, an affected father cannot transmit the disease to his sons, but only to his daughters. All males with the mutation will display the disorder but females are only affected if they are homozygous. This means that in practice X-linked recessive disorders are usually only found in males and rarely seen in females. A feature of X-linked diseases is the frequency with which new mutations are found. In the case of an autosomal recessive disease,

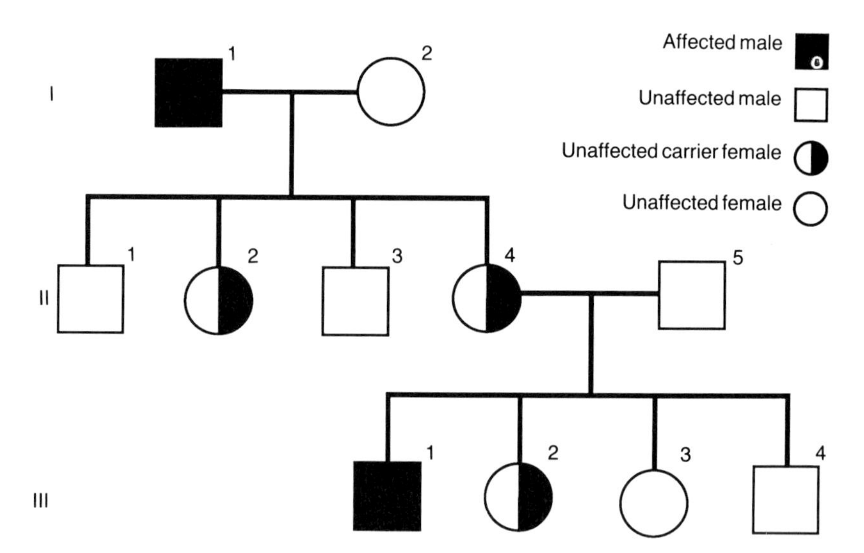

Fig. 4.5 Pedigree of a family displaying inheritance of an X-linked (recessive) disorder. In general, only males are affected, females are either normal or carriers and male-to-male transmission of the disease is not seen.

such as cystic fibrosis, it would be possible, if DNA were available, to trace the mutation back through many generations. In the case of some X-linked diseases the mutation is often in the affected individual or his mother and cannot be traced any further back. For example, Duchenne muscular dystrophy is a muscle defect of early onset with an incidence of 1 in 5000. There is muscular wasting to the extent that the child is wheelchair-bound by 10 years of age and usually dies in his twenties. In Duchenne muscular dystrophy one-third of all cases are due to new mutations. The gene for Duchenne muscular dystrophy was localized to Xp21 in 1983 and in 1987 the gene was cloned. Prenatal diagnosis is now readily available.

In general, in X-linked recessive diseases, only males are affected. Females, since they have two X chromosomes, are either normal or carriers. However, in populations in which the gene defect is common, an affected male and a carrier female may produce affected female offspring. For example, in some populations the glucose-6-phosphate dehydrogenase mutation is sufficiently common for affected females to be occasionally seen. Only a few known genetic disorders are X-linked dominant; one example is hypophosphataemia, which causes vitamin D-resistant rickets.

The increased incidence of mental retardation in males over females seems to be a consequence of X-linked inheritance. A common form of X-linked mental retardation has now been defined that can be diagnosed using a cytogenetic marker on the X chromosome. The incidence of this fragile X syndrome could be as high as 1 in 1000 male births. The fragile site is a constriction at the end of the X chromosome at Xq27. This syndrome cannot be unequivocally assigned to either the X-linked recessive or the X-linked dominant category.

Multifactorial disorders

Disorders arising from both genetic and environmental factors are common and seem especially frequent in congenital malformations. Multifactorial disorders tend to recur in families but do not show the consistent pedigree patterns of single-gene defects. However, in a series of families, a common component may be observed. Studying the aetiology of such diseases can be both practically and intellectually difficult since it is often impossible to differentiate the genetic from the environmental factors. Common congenital malformations with multifactorial inheritance include cleft lip with or without cleft palate, congenital dislocation of the hip, congenital heart disease and neural tube defects. In some cases a disease which is of multifactorial aetiology in the general population can be found as a purely genetic defect in certain families. For example, there is a large Icelandic family in which cleft palate has a purely X-linked inheritance, and a second family which appears to carry spina bifida as an X-linked trait. In the case of the cleft palate family the mutation has been localized, by linkage analysis, to Xq21.3. Work is also in progress to localize the X-linked spina bifida gene. Identification of the mutation which causes the malformations in these two families may give clues as to the aetiology of the same diseases in the general population, where environmental factors also play a part.

The concept of heritability has been put forward to try to untangle the confusion caused by multifactorial inheritance. Heritability is defined as the proportion of the total phenotype variance of a trait that is caused by additive genetic variance or a measure of how large the genetic input is in a given phenotype. One way of estimating this is by comparison of the ratio of the concordance rate of a condition in monozygotic (MZ) and dizygotic (DZ) twin pairs with the incidence in the general population. Monozygotic twins have the same genetic information available to them for development whereas dizygotic twins are equivalent to siblings. For cleft lip with or without cleft palate the concordance in MZ is 0.3 whereas in DZ it is 0.05, and for dislocation of the hip it is 0.4 in MZ and 0.03 in DZ. This and other large family studies group these two disorders as multifactorial and not a consequence of purely environmental factors.

5: Genomic imprinting

The male and female parental contributions to the genome are not fully equivalent. There is increasing evidence that the function of a chromosome may differ depending upon whether it is maternally or paternally derived. For example, it appears that in early development it is mostly paternally derived genes which control the development of the placental tissues whilst maternally derived genes play a more important role in development of the embryo proper. There are a number of examples, of both chromosomal and single-gene diseases, where the phenotype depends upon the parent from which the disease has been inherited. The differences between the maternally and paternally derived chromosomes appear to remain fixed through successive mitotic divisions. This has been termed 'genomic imprinting'. Genomic imprinting must affect a chromosome in a way which survives mitosis but not meiosis. At meiosis the chromosome must be newly imprinted, depending upon the gender of its 'host'.

The actual mechanisms of genomic imprinting are poorly understood at present but it does appear that imprinting occurs at multiple levels; the individual chromosome, the subchromosomal region or the single gene. One of the most popular current theories for the mechanism of imprinting is selective methylation of the genome. In females one of the two X chromosomes in each cell is inactivated so that only genes on one X chromosome are expressed. Which of the two X chromosomes is inactivated in any particular cell is a random process. X inactivation depends, at least in part, on the methylation of CG-rich regions adjacent to a gene on the inactive chromosome. Treatment of cells with a demethylating agent can reactivate these genes. Methylation of the inactive X chromosome is analogous to imprinting, although it affects the entire chromosome rather than parts of it and is not dependent upon the gender of parental origin. The methylation status of a DNA region can be investigated using isoschizomer restriction endonucleases having differing sensitivity to methylation (see Chapter 7). In many cases it has been found that methylation of a gene, particularly in the CG-rich islands associated with the 5′ untranslated regions, correlates with inactivity of that gene. It may be that methylation of a gene inhibits the activity of transcription factors and is an important mechanism in controlling gene function, although it is also possible that methylation is a secondary effect of inactive genes and can only occur when transcription factors are not bound to the gene. There is evidence from transgenic mice that the methylation status of individual chromosomes may vary depending upon the parent of origin and will change as the gene is inherited from male to female and vice versa.

Androgenones and gynogenones

A complete hydatidiform mole has a normal karyotype but is androgenetic, i.e. all of its chromosomes are of paternal origin. The most common mechanism for this is the duplication of one haploid chromosome set so that each of the homologous chromosomes is identical. Occasionally it arises from dispermy and loss of the maternal genetic material. In this case homologous chromosomes, although both paternally derived, will not be identical. A complete hydatidiform molar pregnancy is characterized by aberrant placental development with no development of the embryo proper. The chorionic villi are distended, with multiple cysts, and there is no fetal vascularization. A partial hydatidiform mole is triploid and usually diandric with two paternal sets of chromosomes. In these pregnancies there is molar change in the placenta, and fetal development usually ceases early. When, rarely, the conception is digynic the placenta develops normally, there is a greater chance of fetal survival and occasionally live birth may be achieved.

There are no natural examples of genuine gynogenetic conceptions, although ovarian teratomata have a 46XX chromosome complement, all of which is maternally derived. Frequently they are homozygous for polymorphisms where the 'host' is heterozygous. This suggests that these tumours arise by abnormal development following the first meiotic division. These tumours consist of differentiated ectodermal, mesodermal and endodermal structures but have no placental elements.

These 'experiments of nature' suggest that the paternal genome is particularly important in the development of the extraembryonic tissues whilst the maternal genome has a more important role in embryonic development. These conclusions have been confirmed in experiments with mouse embryos. It is possible, after *in vitro* fertilization, to remove the maternal pronucleus from a mouse embryo and to replace it with a second paternal pronucleus derived from either the same or a different male mouse. Similarly the paternal pronucleus can be exchanged for a second maternal pronucleus. In these cases the gynogenones develop normally until implantation, when the pregnancy fails, whilst the androgenones produce an anembryonic pregnancy with development of only the extraembryonic tissues. These findings have suggested that human anembryonic pregnancy, often referred to clinically as a blighted ovum, may also be androgenetic. However, experiments using DNA fingerprinting to identify the paternal and maternal genetic contributions have shown that these pregnancies do have genetic contributions from both parents.

Genomic imprinting and chromosomal defects

The concept of genomic imprinting suggests that in certain cases a genetic defect will only produce a phenotype if inherited from a particular parent. For example, a chromosomal deletion in a region concerned with placental development may have no effect if inherited maternally, but may cause failure of placental development if inherited paternally. Examples of chromosome deletion syndromes where this seems to apply are the Prader–Willi and Angelman syndromes. The Prader–Willi syndrome is characterized by hypotonia in infancy, developmental delay, obesity and hypogonadism. It is associated with deletions of chromosome 15q11–13. In some cases the deletion is detectable by cytogenetic studies, while in other cases it is submicroscopic and can only be detected by using DNA probes. In individuals who have the Prader–Willi syndrome the deleted chromosome is always paternally derived. The Angelman syndrome is characterized by a happy disposition, mental retardation, ataxic movements, a large mouth and protruding tongue and seizures. Angelman syndrome is also associated with deletions of 15q11–13 but in this case the deleted chromosome is always maternally inherited. It is possible that similar differences in phenotype may be seen with other deletions, depending upon the parental origin. When siblings have the same disorder but have phenotypically normal parents, it is often assumed that this represents an autosomal recessive inheritance. However, it is possible that these may represent chromosome deletions in imprintable regions which have no effect in the parent but, since the imprinting status changes with meiosis, do have an effect in the offspring.

Genomic imprinting and single-gene defects

In certain autosomal recessive conditions there is a difference in the expression, severity or age of onset of the disease, depending upon the gender of the affected parent. The clearest example of the effects of genomic imprinting on a single-gene disease is the hereditary glomus tumour. This rare, benign tumour has an autosomal dominant inheritance but is only seen in individuals who have inherited the disease from their father. The gene is presumably imprinted in the female germ cell line so that it is not expressed in the offspring of affected mothers. The disease might appear to jump a generation when inherited by a female, whose sons would not exhibit the disease but whose grandchildren could do so.

In Huntington disease the age of onset of symptoms is lower in those who inherit the gene from their father than in those who inherit it from their mother. There is a considerable excess of early-onset cases born to affected fathers and a subgroup in which onset is on average 24 years earlier than the age of onset in their father. Late-onset cases have more frequently inherited their disease from their mother. The average

age of onset for the offspring of affected women is 42 years, compared with 33 years for the offspring of affected men. These findings could be explained in terms of imprinting although other explanations would include a role for mitochondrial DNA, which is always maternally derived, or for the X chromosome.

Myotonic dystrophy is another disease in which severity varies with parental origin. The disease is usually characterized by progressive muscular wasting. There is a myotonia and there may be cataracts and gonadal atrophy. Onset of the disease is usually in adult life. There is, however, a rarer, congenital, form of myotonic dystrophy, with severe muscular weakness resulting in intrauterine or neonatal death. Many of those infants born alive die in the first few weeks of life from respiratory disease; a small proportion survive but are severely affected. This congenital form of myotonic dystrophy only occurs in the offspring of affected mothers. It is never seen in cases where the disease has been inherited from the father. Although this could be caused by some interaction between the affected fetus and the intrauterine environment of an affected mother, it is more likely to be an example of the effects of genomic imprinting.

Uniparental disomy

Uniparental disomy is when both of a pair of homologous chromosomes are inherited from the same parent. If the two chromosomes are identical duplications, with the aneuploid event occurring at the second meiotic division, this is termed 'heterodisomy'. If the two are non-identical homologues, with the aneuploid event occurring at the first meiotic division, it is termed 'isodisomy'. The mechanisms of haploid uniparental disomy are not fully understood at present. If it arises only when a gamete with an extra chromosome meets with a gamete with that chromosome missing, it would be a very rare event. It is more likely that it arises by combination of a trisomic gamete with a normal gamete. The cell's selective pressure to eject one of the three homologues may cause the extra chromosome to be lost in early development and in some cases this may leave two homologues from the same gamete. Abnormal karyotypes have been reported in over 40% of oocytes and 5% of spermatozoa. These high rates of aneuploidy in gametes suggest that, if this is its mechanism, uniparental disomy may be fairly common.

There are numerous recognized cases of disomy of the sex chromosomes, 47XXX and 47XXY, since these are easily identified by cytogenetic studies. Diploid isodisomy is very infrequently recognized, since this requires analysis of DNA polymorphisms. There is a reported case of the father-to-son transmission of haemophilia A. This is usually impossible since the male offspring of an affected male inherit only his Y chromosomes and the haemophilia defect is on the X chromosome. In this particular case it was found that the male child had inherited both X and Y chromosomes from his father. The maternal X chromosome was presumably absent in the fertilized ovum or, more probably, lost in early development. There are also cases of cystic fibrosis in which only one parent was a carrier and the child had uniparental disomy for chromosome 7. These cases were identified by the study of DNA polymorphisms around the cystic fibrosis locus. Interestingly in these cases the child was also growth-retarded at birth. Babies with cystic fibrosis are usually of normal birthweight. This possibly resulted from the influence of other, imprinted, genes on chromosome 7. Mice have been bred with uniparental disomy. In mice, uniparental disomy does not usually cause major congenital abnormalities but does cause variations in growth and behaviour. It was found that with uniparental disomy of certain mouse chromosomes there was either overgrowth or growth retardation, depending upon the parental origin of the disomy.

Part 2
Molecular Biology Techniques

6: Isolation of DNA

DNA can be obtained from any living cell with the exception of erythrocytes, which are anucleate. The technique is essentially similar for all cell types. Firstly, the cells are broken open, either mechanically or chemically. Protein is then removed from the DNA by incubation with a proteinase. DNA and RNA are then separated from all other cell constituents by mixing with phenol. Nucleic acids are unique amongst biochemicals in that they do not dissolve in phenol. After centrifugation the phenol separates from the aqueous phase. Proteins and fats remain within the phenol but DNA and RNA are found within the aqueous phase. Any phenol still present can be removed by mixing with chloroform. The phenol dissolves in the chloroform but the nucleic acids remain within the aqueous phase. Once a relatively pure solution of nucleic acids has been made, addition of salt and cold ethanol to bring the ethanol concentration to 66% will cause the DNA to become insoluble and to precipitate. Genomic DNA is of such a high molecular weight that it forms a visible precipitate, with the appearance of a small piece of cotton wool. This can be lifted out of the solution with a glass hook and redissolved in an ethylenediaminetetra-acetic acid (EDTA)/water solution ready for analysis. DNA of smaller molecular weight such as that from bacterial plasmids cannot be hooked out and is collected by centrifugation. Any contaminating RNA can then be removed by treatment with an RNase.

The concentration of DNA in solution can be determined by spectrophotometry. DNA absorbs light at a peak wavelength of 256 nm. A DNA solution of 1 mg/ml has an optical density at 256 nm of 20. Protein absorbs light at a peak wavelength of 280 nm. The ratio of optical densities at 256 and 280 nm is an indication of the purity of the DNA solution.

Sources of DNA

A common source of human DNA is blood. Ideally blood should be taken into an EDTA anticoagulant tube. The EDTA acts both to prevent clotting and to chelate calcium ions away from calcium-dependent DNA endonucleases. Other anticoagulants, however, do produce acceptable results. The cells are lysed in a low-molarity salt solution and the leucocyte nuclei are collected by centrifugation. These are then treated with proteinase, phenol and chloroform and the DNA is precipitated. A single 10 ml blood sample will yield up to 500 μg of DNA, sufficient for over 100 Southern analysis experiments.

DNA can be extracted from tissue samples. The quality of the DNA will depend upon the age of the tissue and its storage conditions. Frequently DNA from tissue samples is partly degraded and is of lower molecular weight than DNA from blood. Tissue should ideally be stored at −70°C although DNA has been successfully isolated from historical pathology specimens in formalin and from Egyptian mummies.

Chorionic villus biopsy is probably, from the point of view of the molecular biologists, the best source of DNA from an unborn child. Chorionic villi are fetal in origin. Chorionic villus biopsy in the first trimester of pregnancy has become available in most areas. It allows antenatal diagnosis early in pregnancy. An affected pregnancy can then be more safely terminated during the first trimester than following amniocentesis in the second trimester. After collection, the chorionic villi are washed under a low-magnification binocular microscope to remove contaminating maternal cells. DNA isolation follows by a method similar to that used for tissue. A single chorionic villus biopsy will yield up to 100 μg of DNA, sufficient for 25 Southern analysis experiments. Amniocentesis has the disadvantages that it is usually performed in the second trimester of pregnancy, it may require several weeks of culture to obtain sufficient DNA for analysis and even then it yields much smaller amounts of DNA than chorionic villus biopsy. Chorionic villus biopsy is, at present, the procedure of choice for obtaining fetal DNA for prenatal diagnosis.

DNA can be extracted from buccal cells either scraped from the inside of the mouth or centrifuged from a mouthwash, from semen, from hair roots and also from Guthrie spots. In these cases the amount of DNA isolated may be too little for conventional analysis but may be analysed by using the polymerase chain reaction.

7: Enzymes used in molecular biology

Restriction endonucleases

An enzyme which removes nucleic acids from either end of a DNA molecule is known as a DNA exonuclease. An enzyme which cuts within the molecule is an endonuclease. Most exonucleases show no specificity for particular bases or sequences of bases. The restriction endonucleases, however, will only cut at sites within specific sequences of DNA (Table 7.1). Restriction endonucleases were discovered when DNA molecules from one subtype of *E. coli* were introduced into another subtype, and were quickly fragmented into smaller, genetically inactive pieces. Many types of bacteria contain a restriction system consisting of a methylase and an endonuclease, both of which recognize the same sequence of bases (restriction site). The methylase acts to methylate residues at the recognition site within the native DNA. The endonuclease will not act at the methylated site but will cut any unmethylated foreign DNA which is introduced into the bacterial cell.

Table 7.1 Some commonly used restriction enzymes and their recognition sequences.

Enzyme	Recognition site	Cut ends	Overhang type
HaeIII	GGCC CCGG	GG | CC CC | GG	No overhang 'Blunt End'
TaqI	TCGA AGCT	T | CGA AGC | T	5′ overhang 'sticky end'
EcoRI	GAATTC CTTAAG	G | AATTC CTTAA | G	5′ overhang 'sticky end'
HindIII	AAGCTT TTCGAA	A | AGCTT TTCGA | A	5′ overhang 'sticky end'
BamHI	GGATCC CCTAGG	G | GATCC CCTAG | G	5′ overhang 'sticky end'
PstI	CTGCAG GACGTC	CTGCA | G G | ACGTC	3′ overhang 'sticky end'
NotI	GCGGCCGC CGCCGGCG	GC | GGCCGC CGCCGG | CG	5′ overhang 'sticky end'

Within bacteria, therefore, the restriction system acts to protect the cell against foreign DNA introduced by plasmids or phage.

The terms restriction endonuclease and restriction enzyme have become almost synonymous since it is the endonucleases rather than the methylases that are used most widely in molecular biology. There are, however, uses for the restriction methylases, which we will come across when discussing complementary DNA (cDNA) library synthesis. A large number of restriction enzymes have now been identified. Some recognize specific groups of four bases and, therefore, cut DNA into many small pieces. Such enzymes are often referred to as four-cutters. Enzymes which recognize groups of six bases (six-cutters) cut less frequently and yield larger fragments, and enzymes which recognize even larger groups cut rarely (rare-cutters) (Fig. 7.1). Restriction sites are frequently palindromic. For example, the enzyme *EcoRI* recognizes the 5′ to 3′ sequence GAATTC; this sequence has the palindromic complementary sequence 3′ to 5′ CTTAAG. Most of the restriction enzymes used routinely in molecular biology cut the DNA within their recognition site, although some enzymes cut the DNA several base pairs away from the recognition site. Those that cut on both strands at the same place

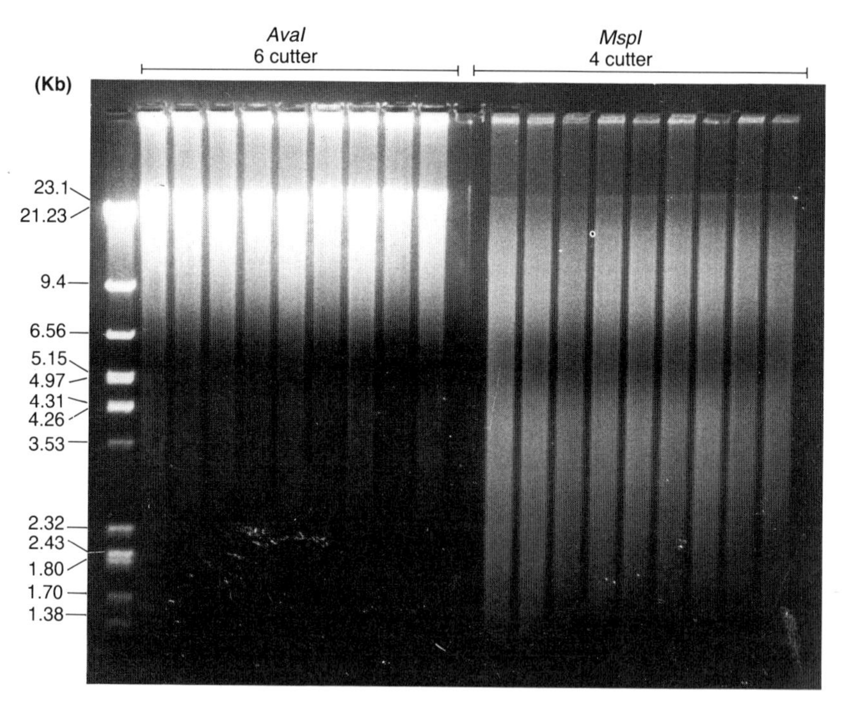

Fig. 7.1 Electrophoresis of genomic DNA digested with the restriction enzymes *AvaI*, which recognizes a sequence of six bases, and *MspI*, which recognizes a sequence of four bases. Since a sequence of six bases occurs less frequently in the genome, the average size of DNA digested by a six-cutter is of higher molecular weight.

produce a blunt-ended cut. Those that cut obliquely leave overhangs of single-stranded DNA. Since these overhangs are complementary and will religate in the presence of a DNA ligase, they are known as cohesive or sticky ends. A restriction endonuclease may cut the DNA to leave either a 5′ to 3′ overhang or a 3′ to 5′ overhang. Restriction enzymes isolated from different bacterial types but which recognize the same DNA sequence are called isoschizomers. Certain isoschizomers, although recognizing the same sequence, are sensitive to methylation of the DNA at different sites. For example, the enzyme *MspI* recognizes the sequence CCGG but will not cut if the first C is methylated. *HpaII* recognizes the same sequence but is sensitive to methylation of the second C. Methylation sensitivity also affects the frequency with which any particular enzyme will cut genomic DNA. If a large number of its recognition sites are appropriately methylated, it will appear to cut less frequently than an isoschizomer which is not methylation-sensitive. Isoschizomers with differing sensitivities to methylation have been used in experiments to identify sites of methylation within the genome.

One of the principal values of restriction enzymes in molecular biology is that they cut DNA molecules into consistent and predictable pieces. For example, digestion of bacteriophage lambda (λ) by a particular restriction enzyme will consistently yield a series of fragments of specific sizes, depending upon the sites of the recognition sequence within the λ genome. Since the entire nucleotide sequence of bacteriophage λ is known, the size of the fragments can be predicted and they may be used as standards to compare with DNA of unknown size. Similarly, digestion of total human DNA with a single restriction enzyme also results in a series of fragments of varying size. The number of fragments is so large that after separation by electrophoresis the total DNA appears as a streak. On closer analysis it will be found that this streak is made up of a large number of individual DNA fragments each corresponding to the distance between two restriction sites.

Other DNA nucleases

There are a number of enzymes with endonuclease activity which are not confined to particular recognition sequences. Deoxyribonuclease (DNase) I hydrolyses both double- and single-stranded DNA at sites adjacent to pyrimidine molecules (C and T). If allowed to digest DNA to completion, it produces a mixture of single nucleotides and small (oligo-) nucleotides. One important use of DNase I is in the production of small nicks in DNA prior to nick translation labelling. S1 DNA nuclease degrades both single-stranded DNA and single-stranded RNA. Double-stranded molecules, DNA : DNA or DNA : RNA are relatively immune from digestion. S1 may be used to remove single-stranded overhangs from double-stranded DNA molecules or in the analysis of DNA : RNA hybrids.

S1 NUCLEASE ANALYSIS OF RNA

S1 nuclease analysis of RNA is a method for identifying the number and position of introns using DNA:RNA hybrids. Double-stranded DNA is denatured to single strands and incubated with the mRNA under conditions which favour DNA:RNA hybrids rather than DNA:DNA hybrids. Treatment with S1 nuclease hydrolyses the single-stranded DNA but the DNA which has hybridized to the RNA is protected from digestion. The sizes of the protected fragments of DNA can be assessed by electrophoresis. DNA from different segments of the gene of interest are used and a map can be constructed of the beginning and end of the coding sequence and the size and position of introns.

DNA polymerases

A DNA polymerase is an enzyme which catalyses the incorporation of single nucleotides into a growing DNA molecule. DNA polymerases require a template of single-stranded DNA and a region of double-stranded DNA to prime their action. They will then add single nucleotides on to the end of the primer in a 5′ to 3′ direction, complementary to the template strand. *E. coli* DNA polymerase I also has 5′ to 3′ and 3′ to 5′ exonuclease activity as well as 5′ to 3′ polymerase activity. This means that it will remove nucleotides as it moves along the template DNA. This feature of DNA polymerase I makes it suitable for radiolabelling of DNA probes by nick translation.

LABELLING DNA PROBES BY NICK TRANSLATION (Fig. 7.2)

In the nick translation reaction double-stranded DNA is treated with limiting amounts of DNase I. The DNase makes small nicks in the double-stranded molecule, removing a few nucleotides at a time. The action of DNA polymerase I is primed by these nicks. Its 5′ to 3′ polymerase activity adds new nucleotides. As the enzyme moves along the template strand and the nicks become filled in, the 5′ to 3′ exonuclease activity removes nucleotides ahead of the enzyme, whilst the polymerase activity adds new ones behind it. If the nucleotides supplied to the enzyme in the reaction mixture are radiolabelled, the resulting product will be radiolabelled and can be used as a hybridization probe.

THE KLENOW FRAGMENT OF DNA POLYMERASE I

The 5′ to 3′ exonuclease activity of DNA polymerase I can often be disadvantageous since it degrades the 5′ end of primers bound to DNA templates and may remove phosphate groups from the 5′ ends of DNA molecules, preventing ligation to other DNA molecules. The domain of DNA polymerase I which has the 5′ to 3′ exonuclease activity can be removed without affecting its polymerase activity. The 3′ to 5′ exoa

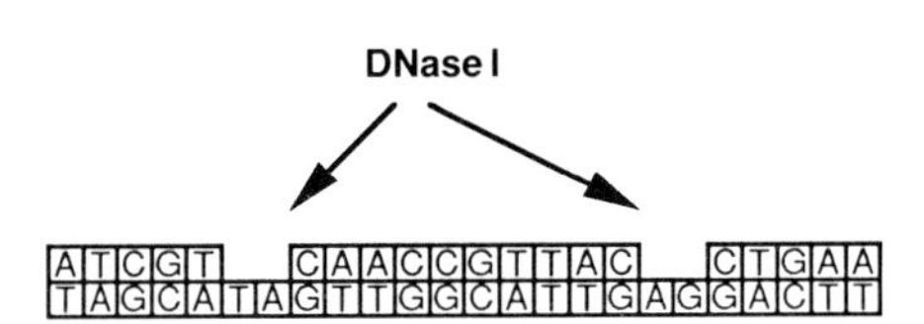

b

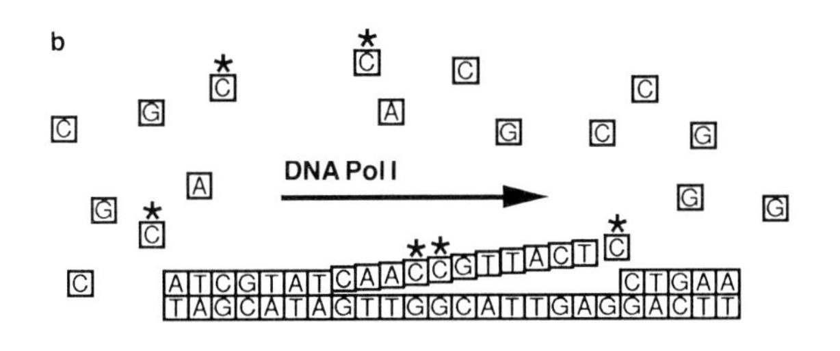

c

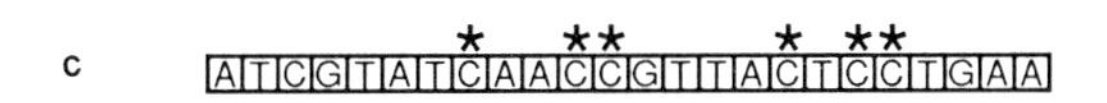

Fig. 7.2 Nick translation. (a) The DNA probe is initially a double-stranded molecule. Treatment with DNase I causes nicks in the molecule. These occur on both strands, but are shown here on only one for simplicity. (b) DNA polymerase I is primed by the regions of double-stranded DNA and synthesizes new complementary DNA. As it moves along the template its 5′ to 3′ exonuclease activity removes nucleotides ahead whilst its polymerase activity adds new nucleotides to the 3′ end of the growing complementary strand. The reaction mix contains at least one radiolabelled nucleotide, in this case dCTP, so the new complementary strand is radiolabelled. (c) Prior to use in hybridization the unincorporated nucleotides are separated from the larger molecular weight probe by column chromatography. The probe is then boiled to denature the double-stranded molecule to single strands and quenched on ice to prevent the single strands from reannealing.

nuclease activity is also unaffected. The original 'large fragment of DNA polymerase' or 'Klenow fragment' was made by chemical cleavage of the 5′ to 3′ exonuclease activity, but the enzyme is now made using recombinant DNA technology.

The Klenow fragment of DNA polymerase I, often known simply as 'Klenow', has a large number of uses in molecular biology. An alternative method of radiolabelling DNA is the random hexanucleotide method.

RANDOM HEXANUCLEOTIDE LABELLING (Fig. 7.3)

In this method of radiolabelling DNA, the DNA is first boiled for several minutes to denature it to single-stranded molecules. It is then quench-cooled to prevent specific reannealing. Random hexanucleotides

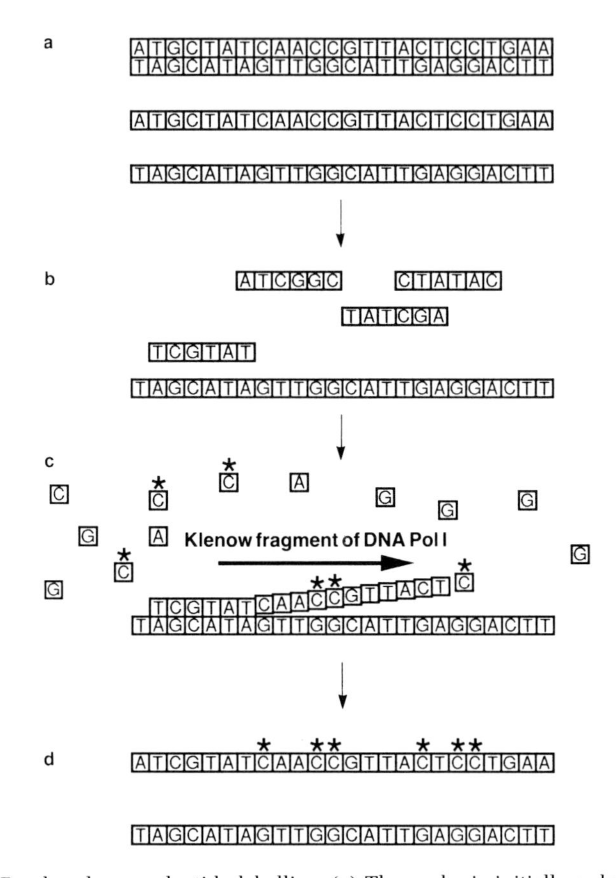

Fig. 7.3 Random hexanucleotide labelling. (a) The probe is initially a double-stranded molecule. It is denatured to single strands by boiling and quenched on ice to prevent the strands from reannealing. (b) The reaction mixture contains random hexanucleotides. Where a hexanucleotide is complementary to one of the probe single strands it anneals. Hexanucleotides will bind to either of the single strands although only one is shown here for simplicity. (c) The Klenow fragment of DNA polymerase I is primed by the regions of double-stranded DNA and synthesizes new complementary DNA adding nucleotides to the 3′ end of the growing complementary strand. The reaction mix contains at least one radiolabelled nucleotide, in this case dCTP, so the new complementary strand is radiolabelled. (d) Prior to use in hybridization the unincorporated nucleotides are separated from the labelled DNA by column chromatography. The probe is boiled to denature the double-stranded molecule to single strands and is then quenched on ice to prevent the single strands from reannealing.

are, as the name suggests, six-base single-stranded DNA molecules with a random base composition. When added to the template single-stranded DNA, they anneal at various sites where they find complementary sequences. Klenow is primed by the short stretches of double-stranded DNA where hexanucleotides are annealed to the template and adds new nucleotides in a 5′ to 3′ direction. If the nucleotides supplied to the enzyme in the reaction mixture are radiolabelled, the resulting product will be radiolabelled and can be used as a hybridization probe.

Klenow can also be used to add radiolabelled nucleotides to the end of DNA with recessed 3′ ends. Many restriction enzymes leave a recessed 3′ end after digestion. The enzyme is primed by the double-stranded DNA and adds new nucleotides to the 3′ end in the 5′ to 3′ direction. Since the enzyme will replace nucleotides complementary to the opposite sequence for radiolabelling, it is important that the correct radiolabelled nucleotide is in the reaction mix. In many laboratories only one radiolabelled nucleotide is routinely used, e.g. deoxycytosine triphosphate (dCTP). However, to end-label a fragment of DNA produced by digestion with *EcoRI*, which leaves a 5′ overhang of AATT, either radiolabelled dATP or dTTP will need to be in the reaction mix, since only these nucleotides will be incorporated.

The Klenow fragment of DNA polymerase I may also be used for second-strand synthesis of cDNA (this is discussed in Chapter 10). It has been used in sequencing reactions and for the polymerase chain reaction, although in these two latter cases more efficient DNA polymerases have become available.

BACTERIOPHAGE DNA POLYMERASES

Phage-encoded DNA polymerases have also been isolated from infected bacteria. Bacteriophage T7 DNA polymerase, for example, is the most processive of DNA polymerases known. This means that it adds new nucleotides at a very fast rate, making it ideal for use in sequencing reactions using the Sanger dideoxy-chain termination method (discussed in Chapter 11). It also has a powerful 3′ to 5′ exonuclease activity but, as with the Klenow fragment, this can be removed chemically. A completely exonuclease-free version of bacteriophage T7 DNA polymerase, for use in DNA sequencing, has been made by genetic engineering.

THERMOSTABLE DNA POLYMERASES

A thermostable DNA polymerase was originally isolated from the extreme thermophile *Thermus aquaticus*. This enzyme, *Taq* polymerase, is most efficient at temperatures between 70 and 80°C. The enzyme is 10 times less efficient at 37°C. Similar enzymes have since been isolated from other thermophilic organisms and a *Taq* polymerase made by genetic engineering has also become available. Although they may also be used for cDNA synthesis (see Chapter 10) and DNA sequencing, the principal use of thermostable DNA polymerases is in the polymerase chain reaction (PCR). Since the enzyme is able to withstand high temperatures, it is possible to denature DNA, reduce the temperature to anneal a primer to the single strand and increase the temperature again to allow the polymerase to synthesize a new complementary second strand without significantly reducing the activity of the enzyme. For full details of PCR see Chapter 9.

REVERSE TRANSCRIPTASE

Reverse transcriptase catalyses the transcription of RNA into DNA. Unlike all other DNA polymerases, it has 5′ to 3′ polymerase activity. The enzyme is a product of retrovirus gene expression and allows the RNA genome of the retrovirus to be transcribed into DNA, which can express the genes needed to produce new virions. In molecular biology it is used to copy mRNA to make complementary DNA (cDNA) (see Chapter 9). Two types of reverse transcriptase are available, purified from avian myeloblastosis virus (AMV) and the Moloney murine leukaemia virus (M-MLV). AMV reverse transcriptase has the great disadvantage that it has RNase activity, which can cause problems when transcribing long mRNA molecules. It does, however, work at a higher temperature (42°C; the body temperature of chickens). RNA molecules with a complex secondary structure may be copied more efficiently at a higher temperature since this helps to keep them denatured.

DNA-dependent RNA polymerases

There are a number of DNA-dependent RNA polymerases: those in common use including bacteriophage SP6, T3 and T7 RNA polymerase. They catalyse the transcription of double-stranded DNA into complementary RNA. Each recognizes a specific DNA sequence called a promoter sequence and begins transcription immediately downstream (3′) of the promoter. They can be used *in vitro* to produce the RNA products of cloned DNA molecules, using special expression cloning vectors.

Other DNA-modifying enzymes

The enzymes used to join DNA together are known as DNA ligases. The most commonly used is bacteriophage T4 DNA ligase, which was isolated from *E. coli* infected with bacteriophage T4. T4 DNA ligase catalyses the covalent joining of the 3′ hydroxyl group and 5′ phosphate terminus of DNA molecules. It works efficiently when joining molecules that have been digested with the same restriction endonuclease and so have compatible cohesive ends. It will also ligate DNA that has blunt ends, although the reaction is slower. To prevent DNA molecules from ligating, the 5′ phosphate can be removed, using alkaline phosphatase. DNA molecules that do not have the required 5′ phosphate can be phosphorylated using a T4 polynucleotide kinase, which catalyses the transfer of γ-phosphate from ATP on to the 5′ terminus of the DNA. The 5′ phosphorylating activity of T4 polynucleotide kinase is also used in the radiolabelling of single-stranded DNA molecules, such as oligonucleotide probes. The enzyme catalyses the transfer of a radio-labelled γ-phosphate from ATP on to the 5′ terminus of the DNA molecule. To radiolabel the 3′ end of double- or single-stranded DNA, terminal deoxynucleotide transferase is used. This enzyme adds deoxynucleotide bases on to the 3′ end of a DNA molecule. If the bases in the reaction mix are radiolabelled, this will label the molecule. This enzyme can also be used to add a tail of nucleotides on to molecules, which is useful in a variety of cloning strategies.

8: Analysis of DNA

Electrophoresis

Once DNA, whether genomic or from bacterial phage or plasmid, has been cut into fragments using a restriction enzyme, the fragments may be separated using electrophoresis. Various media are available for the electrophoresis of DNA, the most widely used being agarose and polyacrylamide. Agarose is a gelling polysaccharide purified from agar. Agar is an extract of various seaweeds. To make a gel for electrophoresis, agarose is dissolved by boiling in an electrolyte solution and then poured into a mould. After setting the gel has the appearance of a thick jelly. Agarose may be used to separate DNA fragments from about 50 000 to 200 base pairs. The rate of DNA migration through an agarose gel is a function of its molecular size and the agarose concentration. To resolve DNA fragments between 1 and 10 kb in size, 0.8% gels are used. Smaller fragments may be resolved on 2% agarose gels. Fragments larger than 10 kb may be resolved on 0.4% gels. Such low-percentage agarose gels are very fragile and require careful handling. For ease of handling, a low-percentage gel may be made within a supporting gel of higher agarose concentration. The voltage applied to the gel will determine the rate at which the DNA fragments separate, although care needs to be taken to ensure that the heat generated by the current passing through the gel does not cause it to melt. The addition of a dye of known molecular size, usually bromophenol blue, allows the progress of the gel to be assessed.

Once the DNA has been separated, it can be visualized under ultraviolet light after staining with ethidium bromide. Ethidium bromide intercollates with DNA and RNA and once complexed will fluoresce in the visible range when illuminated by ultraviolet light. Uncomplexed ethidium bromide fluoresces to a much smaller degree but may contribute high levels of background light and can be washed out of a gel if need be (Figs 8.1 & 8.2).

Polyacrylamide gels allow separation of smaller DNA fragments to as little as a few base pairs. A polyacrylamide gel is made by cross-linking acrylamide molecules in a reaction performed just before the gel is poured between two glass plates sealed on three sides. Whereas an agarose gel is several millimetres thick and is run horizontally, a typical polyacrylamide gel used for DNA analysis is less than a milli-

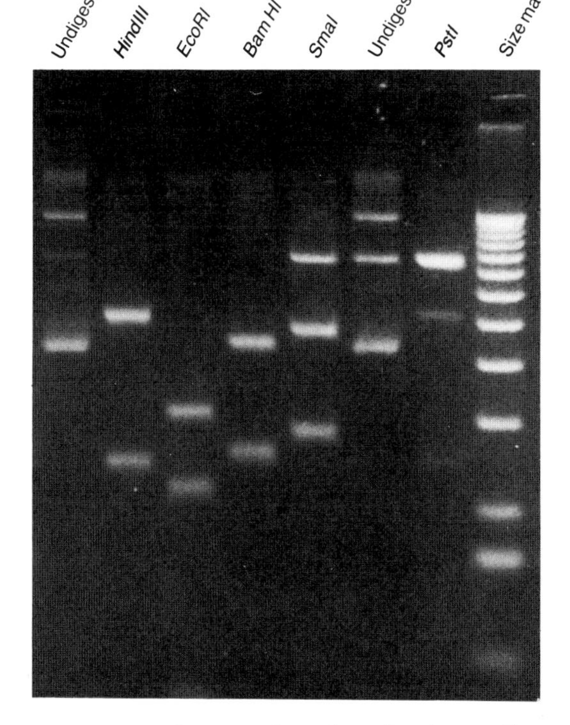

Fig. 8.1 Restriction digests of various plasmids analysed on a 1% agarose gel stained with ethidium bromide. Bands present in the lanes containing undigested DNA represent varying molecular configurations of the circular plasmid. In each case the restriction enzyme recognizes two sites within the plasmid, giving two bands, except Sma1 which recognizes three sites.

metre thick and is run vertically. For some types of analysis, such as in DNA sequencing, the formation of DNA secondary structures can interfere with accurate separation. Urea has been found to be an effective denaturant and can be added to the gel to ensure that the DNA remains in the linear single-stranded form. DNA in a polyacrylamide gel can be stained with ethidium bromide, or transferred to a filter for analysis by hybridization (see below), or the DNA may be radiolabelled before running the gel and the separated fragments identified by direct autoradiography of the dried gel.

PULSED-FIELD ELECTROPHORESIS

The separation of very large DNA molecules, greater than 20 kb, is difficult using conventional agarose gel electrophoresis. The technique of pulsed-field electrophoresis does allow large molecule separation in agarose gels. Rather than the application of a voltage difference from one end of the gel to the other, electrodes are placed both at the ends and around the sides of the gel. The current passes through the gel

Fig. 8.2 Genomic DNA, digested by the restriction enzyme *HincII*, analysed on a 0.8% agarose gel stained with ethidium bromide. Size markers are a restriction digest of the bacteriophage λ.

obliquely and is alternately switched from right oblique to left oblique. With each change in the direction of the current the DNA molecules change direction and configuration before they are able to pass through the agarose. Larger molecules do this less readily than smaller molecules. Smaller molecules begin moving in the gel before larger molecules have had time to change their orientation, and, therefore, migrate through the gel more rapidly.

The earliest pulsed-field electrophoresis apparatus did not produce an overall homogeneous electrical field. This meant that the DNA ran in a curve, tracking away from the midline. More recent apparatus has electrodes arranged hexagonally, or in some other symmetrical pattern, and the pulse pattern is controlled by a microprocessor. This produces straight DNA lanes similar to those seen with conventional electrophoresis. Great care needs to be taken in the preparation of DNA for pulsed-field electrophoresis. Normal preparation techniques usually produce DNA of a molecular weight below 200 kb, whereas pulsed-field electrophoresis aims to separate DNA molecules of between 100 and 5000 kb. To maintain the DNA at as large a molecular weight as possible, the source cells are suspended in an agarose block. This is treated with proteinase K to denature the cellular protein. The proteinase K is then denatured itself, using a proteinase inhibitor, before the block is incubated with a rare-cutter restriction enzyme such as *NotI* or *MluI*. The agarose blocks are then dropped directly into the wells of the gel. Both the gel density and the pulse conditions can be adjusted to

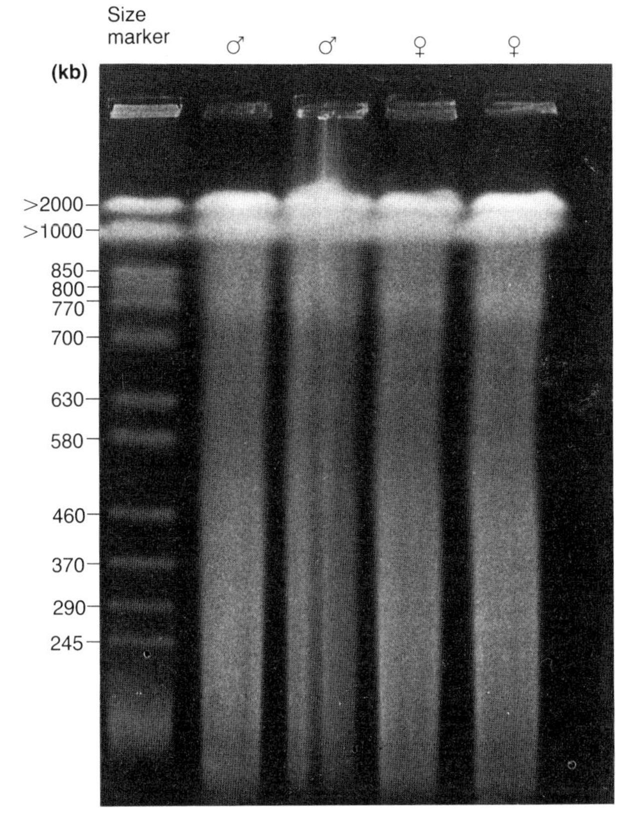

Fig. 8.3 Genomic DNA, prepared from lymphocytes and digested with the rare-cutter restriction enzyme *ClaI*, analysed by pulsed-field electrophoresis. Size markers are whole yeast chromosomes.

separate molecules of different sizes. Increasing the pulse times or decreasing the gel density will increase the size of molecules that can be separated (Fig. 8.3).

Southern blotting

Southern blotting, developed in Edinburgh by Professor E.M. Southern, is a technique which allows specific regions to be identified from DNA that has been fragmented by a restriction enzyme (Fig. 8.4). The DNA is separated using agarose gel electrophoresis. After staining with ethidium the gel is photographed under ultraviolet light. It is then treated with sodium hydroxide. This denatures the DNA and breaks the bonds between its two strands. The gel is then neutralized in a suitable buffer. In a large tray, a thick piece of filter paper is laid over a raised support with its ends dipping in a high-salt solution. The gel is placed on to this filter paper wick on the raised support and a special nylon membrane is placed upon the gel. On top of this is placed a stack of paper towels.

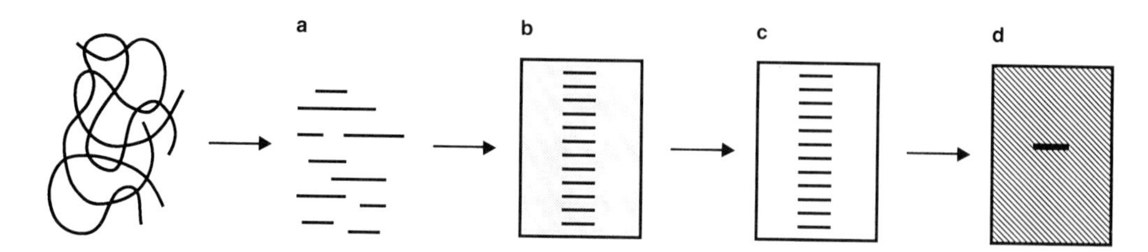

Fig. 8.4 Southern analysis. (a) High molecular weight DNA is digested by a restriction enzyme producing a series of smaller molecules each being the distance between recognition sites. (b) The molecules are separated by electrophoresis. Small molecules travel further down the gel than larger molecules. (c) The DNA is transferred to a nylon membrane by Southern blotting. (d) The Southern blot is hybridized to a radiolabelled DNA probe which detects only specific restriction fragments. The size of the detected fragment can be determined by autoradiography.

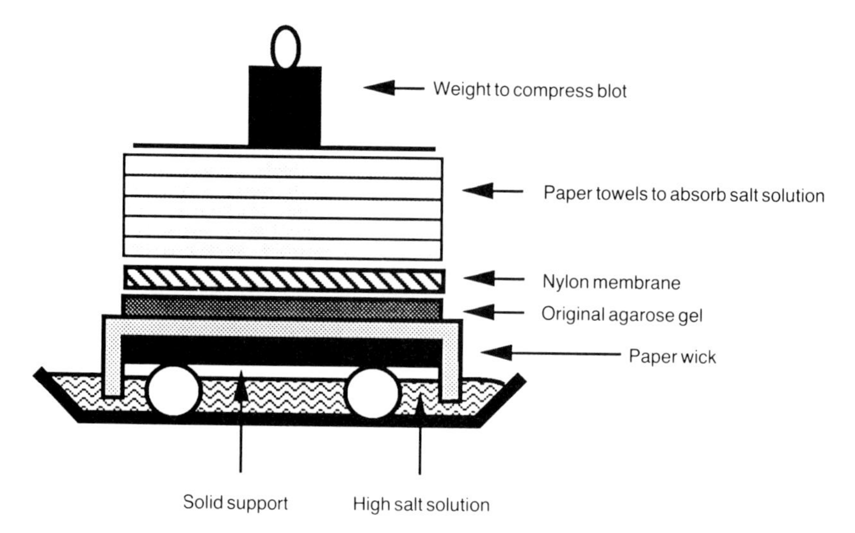

Fig. 8.5 Southern blot. The high-salt solution is drawn up the wick by capillary action and through the gel to the membrane. DNA in the agarose gel is drawn with it through to the nylon membrane. The DNA binds to the membrane electrostatically and can later be fixed permanently either by baking or by cross-linking under ultraviolet light.

Finally all of the layers are kept compressed by a weight (Fig. 8.5). The high-salt solution rises up through the paper wick and then through the gel by capillary action, taking the denatured DNA with it. The DNA is carried through to the nitrocellulose membrane, to which it binds. Thus the filter becomes a replica of the original gel. Once transfer is complete the DNA may be irreversibly bound to the filter by baking or by exposure to ultraviolet light. The use of vacuum-mediated transfer can reduce the blotting time significantly. Pulsed-field electrophoresis gels are pretreated with ultraviolet light to break up the large DNA molecules before Southern blotting. Traditional blotting of pulsed-field gels may take several days rather than the few hours taken for normal gels.

Hybridization

Once a Southern blot has been made it can be hybridized to a radiolabelled DNA probe specific for the gene or region of DNA under study. The filter is first prehybridized for several hours in a solution containing salmon sperm DNA and other substances which block non-specific DNA binding sites. Hybridization to the radiolabelled DNA probe is then performed in a similar solution. It is particularly important to adjust the hybridization conditions so that annealing of complementary sequences takes place but non-specific hybridization is minimized. Increasing the temperature of hybridization or decreasing the ionic strength decreases the annealing of incompletely matched sequences. This is termed 'stringency'. Under conditions of high stringency, i.e. a high temperature and low salt concentration, only perfectly matched sequences will anneal. Under conditions of lower stringency, mismatched sequences will also anneal and there will be an increase in the background, non-specific, hybridization. Of course if the temperature is too high even perfectly matched sequences will not hybridize to each other. Selection of the correct hybridization conditions is less important when large DNA probes, consisting of hundreds or thousands of bases, are used than when small, oligonucleotide, molecules of only a few bases are used. With smaller probes the temperature at which a perfectly matched hybrid would denature to single strands, the melting temperature (Tm), can be calculated. Hybridization is then performed at a temperature a few degrees below this.

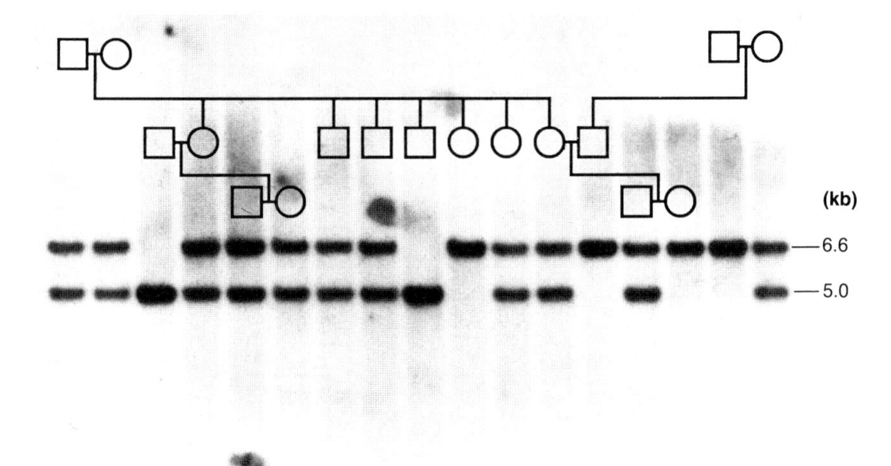

Fig. 8.6 Southern analysis demonstrating a restriction fragment length polymorphism. The agarose gel shown in Fig. 8.2 has been Southern blotted and hybridized to the probe D12S51, which detects a polymorphism with the enzyme *HincII*. Presence of the polymorphic restriction site gives rise to the 5.0 kb band and absence of the site gives the 6.6 kb band. Heterozygotes inherit both bands.

After several hours in contact with the filter, unbound radiolabelled probe is washed off. Washing in solutions of higher or lower stringency can be used to remove non-specifically bound probe or to conserve binding where the probe is not perfectly complementary. The location of bound probe on the filter can then be identified by autoradiography. To determine the size of restriction fragment bearing the sequence of interest, comparison can be made with standards visible either on the photograph of the agarose gel or, if radiolabelled markers were used, on the autoradiograph itself (Fig. 8.6).

9: Polymerase chain reaction

The polymerase chain reaction (PCR) is a powerful technique for amplification of specific DNA or RNA fragments (Fig. 9.1). DNA polymerases will not directly copy a single-stranded DNA molecule unless primed by a stretch of complementary DNA bound at its 5′ end. The PCR takes advantage of this by providing primers which are specific for the region of DNA to be amplified. Two primers are needed, one on each complementary strand, at each 5′ end of the region to be amplified. The reaction mix also contains all four single nucleotides and a DNA polymerase. The DNA is first heated to 95°C which breaks the bonds between the two complementary strands, forming single-stranded DNA. The reaction is then cooled to a temperature at which the specific primers will anneal to the DNA at each 5′ end. The primers in the reaction are in great excess over the template DNA. Template/primer hybridization is favoured over reannealing of the two template strands. The temperature is changed again to the optimal for the DNA polymerase. The DNA polymerase synthesizes a new strand of DNA complementary to the template, beginning at the primer site. The number of molecules of the specific region of DNA has now doubled. Repeating the temperature cycle denatures the newly formed double-stranded DNA, allows the primers to anneal to the single-strand templates and again doubles the amount of specific DNA present. If the temperature cycle is repeated 20 or 30 times, the specific region of DNA, originally present in nanogram quantities, can be amplified to microgram quantities, sufficient to be seen on an ethidium bromide-stained agarose gel.

When PCR was first described, the enzyme used was the Klenow fragment of DNA polymerase I. The optimal temperature for this enzyme is lower than the optimal annealing temperture for most primers and the enzyme is destroyed when the reaction is heated to 95°C. Klenow therefore had to be added to the reaction after each cycle. The discovery of DNA polymerases in microorganisms which live in hot springs has resolved this problem. *Taq* polymerase, from the bacterium *Thermus aquaticus*, is now routinely used. This enzyme has its optimal activity at 72°C and remains active after being heated to 95°C. Other heat-stable DNA polymerases, isolated from other organisms living at higher temperatures, are now becoming available. Since these DNA polymerases are heat-stable, they can be added at the beginning of the experiment and

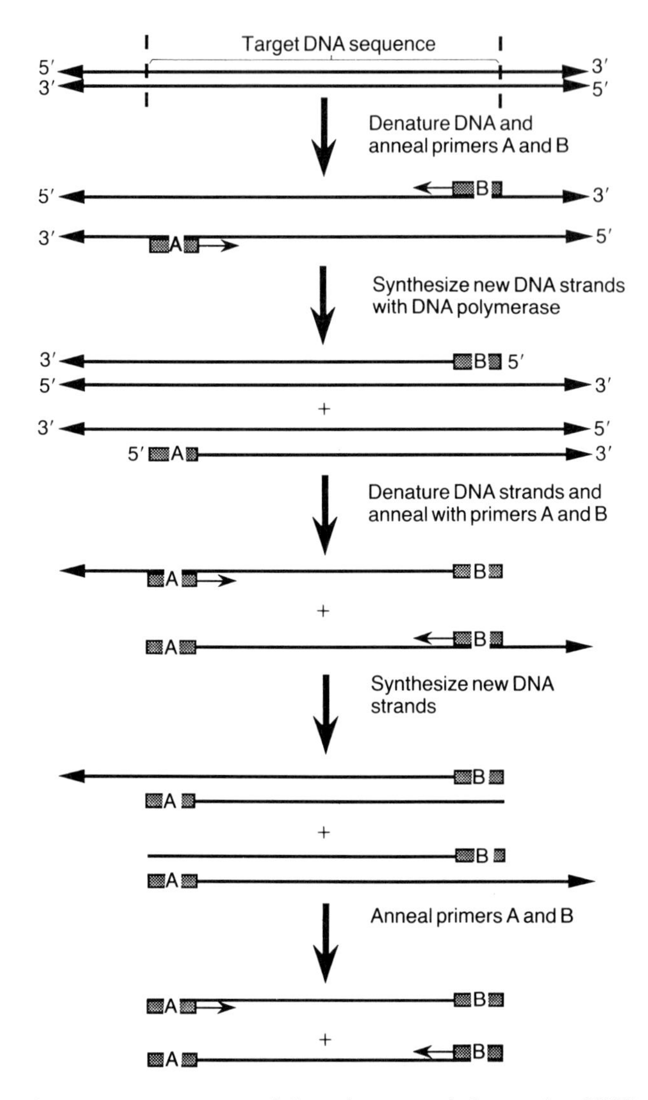

Fig. 9.1 Schematic representation of the polymerase chain reaction (PCR). Subsequent cycles denaturing, annealing and synthesis yield an exponential increase in the amount of PCR product.

do not need to be replenished. The polymerase chain reaction can therefore be automated and various designs of cycling heat blocks are now available commercially.

Oligonucleotide synthesis

Oligonucleotides are synthesized chemically. A number of commercial machines are available but they all use similar principles. Synthesis begins with the most 3′ base, which is bound to a supporting column. To the column are added reagents which contain the next base. This base has an activated 3′ phosphate but its 5′ hydroxyl group is capped by dimethyl trityl. This means that one new nucleotide can link to the 5′ hydroxyl of the base on the column but that no further bases can join on to the 5′ end of the new base. All of the reagents are then washed out and the dimethyl trityl is then removed using an acid. The column now supports a dimer whose 5′ hydroxyl is uncapped. The third base is added, again with an activated 3′ phosphate and a capped 5′ hydroxyl group. One base binds to each of the dimers, producing a trimer. The reagents are washed out again and the cap removed from the trimer, using acid. This cycle is repeated, each time adding a single base until the full oligonucleotide has been synthesized. It can then be removed from the column and made ready for use. Oligonucleotides made in this way are single-stranded DNA molecules.

Oligonucleotide primer design

To perform amplification of DNA using PCR requires knowledge of the nucleic acid sequence of the region at either end of the fragment. Primers are usually designed to be about 20 base pairs in length. Smaller primers increase the risk that other regions in the genome will be homologous. This will cause non-specific priming, resulting in amplification of unwanted regions and depletion of primers and nucleotides. Larger primers are more expensive to synthesize.

Ideally the two primers should have the same annealing temperature. The annealing temperature can be estimated from the formula:

$$T = 2(A + T) + 4(G + C) - 5(°C)$$

The exact ideal annealing temperature will vary depending upon the actual base composition and order, but the formula gives a good guide for practical purposes. Primers are usually designed to be of the same length and to have the same (A + T) to (G + C) ratio, but it is possible to design two primers with similar annealing temperatures having different lengths. From the formula it can be seen that the longer primer would usually need to have more A or T in its composition to have a similar annealing temperature. The annealing temperature is usually 5°C below the true melting temperature (*Tm*) of the amplification primers. The *Tm* is the temperature at which the primer/template hybrid is in equilibrium between annealing and denaturing. A temperature 5°C below the *Tm* is favourable towards annealing but is still high enough to retain specificity. Increasing the annealing temperature should enhance discrimination against incorrect, and therefore not fully complementary, annealing of primer to template. It is particularly important that the temperature of the reaction is not allowed to drop after the first denaturation step such that non-specific priming takes place. Once a non-specific product is amplified, it will have regions at either end complementary to the primers and will be amplified as efficiently as the specific fragment.

Reaction conditions

The time needed for the DNA polymerase to extend from the primer to form a complete complementary strand depends upon the length and concentration of the target sequence and the extension temperature. It is estimated that the rate of incorporation of nucleotides into the growing complementary molecule is up to 100 per second. An extension time of 1 minute is suitable for most applications, although this may need to be increased if the concentration of template is very low or when the concentration of target exceeds the concentration of enzyme in later cycles.

A common cause of PCR failure is incomplete denaturation of the template. Although DNA takes only a few seconds to denature at its strand-separation temperature, allowance needs to be made for the delay between the rise in temperature in the heating block or water bath and the rise in temperature within the reaction tube itself. Some PCR machines have a reaction tube supplied containing the temperature sensor, so that their cycling temperatures are controlled by the temperature within the tube rather than the temperature of the block or bath. To shorten a denaturation step may allow the partly denatured DNA to snap back and hence reduce the overall efficiency. Too long a denaturation step leads to unnecessary loss of enzyme activity. The half life of *Taq* polymerase is over 2 hours at 92°C but only 5 minutes at 97.5°C.

The number of PCR cycles needed to produce adequate amplification of a target sequence will depend upon the starting concentration of the target when all of the other factors are optimized. If there are more than 10 000 target molecules, 30 cycles are usually adequate. With 1000 target molecules, 40 cycles may be needed. When amplification of a sequence from a single cell is required, more than 40 cycles may be needed. As the number of PCR cycles increases, a plateau effect is seen where the rate of amplification is no longer exponential. The plateau seems to occur with the accumulation of about 1 pmol of product. Exhaustion of the nucleotides or of the enzyme will also slow the reaction. Once the plateau phase has been reached for the target sequence and its rate of amplification slows, there is a danger that non-specific products will then become preferentially amplified. The number of cycles should therefore be optimized to ensure adequate amplification of the specific target without the accumulation of background products.

The enzyme requirements may vary with individual target sequences or primer combinations. The enzyme itself is dependent upon magnesium for its function, and the magnesium concentration may also affect strand dissociation and primer annealing to the template. There is no certain way to determine the optimal enzyme and magnesium conditions for any individual reaction and they can often only be found by experimentation. The four single nucleotides (dNTPs) need to be in the reaction mix in equimolar concentrations. If one dNTP is present at a higher concentration, this increases the risk that misincorporation errors will take place during the extension phase. It also appears that too high an overall concentration of dNTPs can increase the risk of mispriming at non-target sites and the likelihood of extending from misincorporated nucleotides.

PCR is usually performed to amplify fragments of less than 2 kb in size. Reaction products are analysed by gel electrophoresis. To separate fragments which are under 1 kb in size, high-density gels of 3 or 4% agarose are used (Fig. 9.2). At these concentrations standard agarose is highly viscous. Better results are often obtained with the use of a low-gelling and melting-temperature agarose. The low viscosity makes pouring of gels easier and at high concentrations fragments of only a few base pairs in length can be differentiated. Alternatively a high-percentage polyacrylamide gel can be used.

The extreme sensitivity of PCR makes it essential to avoid contamination of reactions with template from other sources. Reagents and pipettes can easily become contaminated with the products of previous reactions. This is particularly a danger in a laboratory which regularly amplifies a specific fragment. This can be minimized by reserving specific pipettes and bench areas for setting up PCR and others for analysis of the products. Positive-displacement pipettes, with disposable barrels, may also be used, although these are somewhat expensive. Contamination with genomic DNA, from the operator or from the reagents, may also be a problem, especially when the concentration of genuine template is very low. Adoption of a strict aseptic technique as used in cell culture work can help to minimize this problem. Some groups who amplify Y chromosome-specific regions from single cells

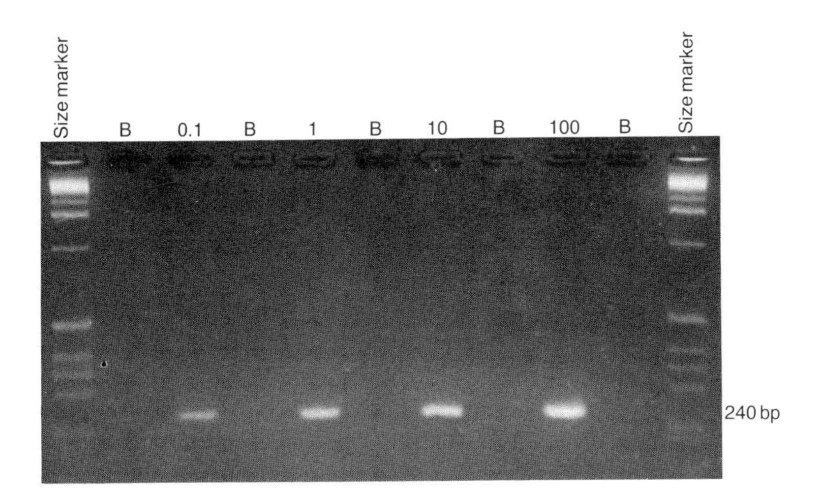

Fig. 9.2 Amplification products of PCR analysed on an ethidium-stained, 3% agarose gel. The high-percentage gel is used to achieve good resolution of small PCR products, in this case 240 bp. B = blank, and DNA quantities are given in pg.

allow only female operators to set up the PCR, hence minimizing the risk of contamination by male genomic DNA.

Polymerase chain reaction is finding an increasing number of applications in molecular biology. It can be used to identify the presence of a specific sequence, such as a viral DNA molecule in a clinical sample (see Chapter 17). It has been used to amplify ancient DNA from, for example, Egyptian mummies, and not such ancient DNA from the Guthrie spots of neonates who have died. DNA sequencing directly from PCR products is now possible, using double-stranded DNA as a template. It is also possible to generate specific single-stranded DNA by PCR. If the ratio of the two primers is adjusted so that one is depleted early in the number of cycles, single-stranded DNA continues to be made from the remaining primer, although the amplification is no longer exponential because the concentration of template does not increase. If a sufficient number of cycles is used, enough single-stranded DNA will accumulate to provide a sequencing template. PCR can be used to amplify RNA, either *in vitro* or *in situ*, and it can be used for direct gene cloning (see Chapter 10).

10: Cloning genes whose protein products are known

The cloning of specific genes is made easier if something is known about its protein product. If the amino acid sequence of the protein is known, it may be used to make a guess at the corresponding nucleic acid sequence. From this, small single-stranded oligonucleotide probes can be constructed and used to screen an appropriate cDNA library. A cDNA library is a DNA representation of all of the mRNA isolated from a particular cell or tissue. The mRNA is isolated and a corresponding DNA molecule is synthesized using reverse transcriptase. The DNA molecules are introduced into a suitable vector which allows them to be replicated. The clone carrying the cDNA corresponding to the gene of interest may be identified by hybridization with the oligonucleotide probe. If the protein has been isolated but the amino acid sequence is not known, antibodies can be raised against the protein. Certain vectors are able to express protein from the cDNA cloned into them. They will produce the protein product of the corresponding gene. A cDNA library cloned into an expression vector can be screened, using the antibodies to identify the clone of interest. More recently techniques have been developed to allow the direct cloning of specific cDNAs from mRNA without the use of libraries, using the polymerase chain reaction.

Plasmid vectors

A vector is a term coined by molecular biologists to describe a DNA molecule that can be used to carry foreign DNA. The vector complete with foreign DNA can be inserted into a host cell. As the host cells grow, the vector and its foreign 'insert' are also amplified. The first vectors to be developed, in the early 1970s, were based on bacterial plasmids. Plasmids are circular DNA molecules which coexist with the bacterial chromosomal DNA within the cytoplasm. They carry functional genes coding, for example, antibiotic resistance and can be passed from one bacterial cell to another (Fig. 10.1). Some plasmids exist in only a single or small copy number, while others multiply in higher numbers within the bacterium. In early experiments it was found that the plasmid pSC101 could be linearized using the restriction enzyme *EcoRI.* A foreign DNA molecule, also digested with *EcoRI,* had ends which were complementary to the ends of the plasmid. When the cut plasmid and

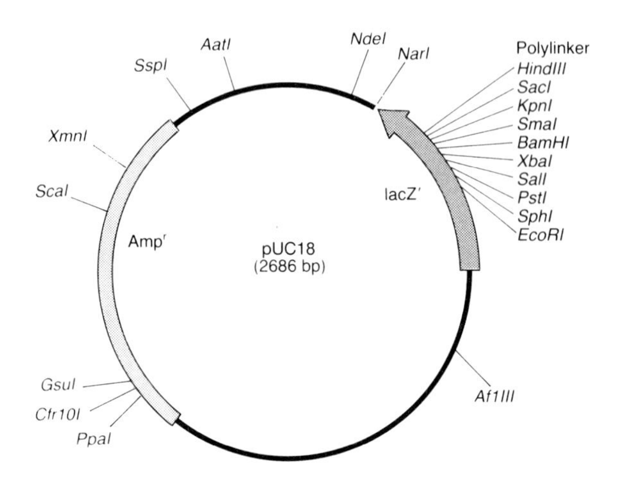

Fig. 10.1 The plasmid cloning vector, pUC18. Important features of this commonly used vector include: the polylinker, which contains a selection of restriction enzymes that cut only once within the vector, ampicillin-resistance selection system; and the lacZ′ gene which allows colour selection of recombinants.

foreign DNA were mixed with the enzyme DNA ligase, the plasmid recircularized and some of the plasmid molecules now contained the foreign DNA as an insert. When the plasmid was introduced into host *E. coli* bacteria the plasmid, together with the foreign DNA, could be replicated many-fold.

There are now a large number of types of plasmid available to the molecular biologist, with various different features built into them. Plasmid vectors can accommodate foreign DNA up to about 10 kb in size. Plasmid vectors carry a gene for antibiotic resistance. This enables bacterial cells which contain the plasmid to be selected from those that do not. In some cases there is a second antibiotic resistance gene, which is interrupted if the plasmid has foreign DNA cloned into it. This allows selection of plasmids carrying foreign DNA, i.e. recombinants, from non-recombinants. Bacteria containing non-recombinants will grow in the presence of both antibiotics, whereas those containing recombinants will grow only in the presence of one. This method of identifying recombinants is somewhat unsatisfactory, since it is the recombinants which do not grow in the selecting medium. As an improvement, plasmid vectors have been designed to allow colour selection. These plasmids contain the gene β-galactosidase, which is interrupted when foreign DNA is inserted into the plasmid. Colonies containing the plasmid are grown on a medium containing the substrate, X-gal (5-bromo-4-chloro-3-indolyl-β-D-galactopyranoside) and IPTG (isopropyl-β-D-thiogalactopyranoside). IPTG induces expression of the β-galactosidase gene, which reduces the X-gal. This forms a blue colour. Non-recombinant colonies are therefore blue. Recombinant plasmids cannot express β-galactosidase and therefore form white colonies.

The earlier plasmids were limited in the choice of enzyme sites for cloning. Foreign DNA could only be cloned into single sites which occurred naturally in the plasmid and did not interrupt genes with the essential functions of the plasmid. More recently vectors have been designed which contain a 'polylinker'. This is an artificially introduced region of plasmid which contains a series of recognition sites for both common and rare-cutter restriction enzymes which do not occur elsewhere in the plasmid. This greatly increases the chance that a suitable restriction enzyme can be found for cloning purposes. When cloning into a single site, it is possible that the insert will be ligated into the vector in either orientation. The 'sense' strand may become inserted in either the 3′ to 5′ or the 5′ to 3′ direction. If, however, the vector is cut with two of the polylinker enzymes, the resulting linear molecule will have different sticky ends. An insert flanked by the same two sites will clone into the vector in only one orientation. This is the technique of 'forced cloning'.

To ligate foreign DNA molecules into plasmid vectors, both the DNA molecule and the vector need to be cut with appropriate restriction enzymes (Fig. 10.2). The two are then brought together in a reaction with a DNA ligase. The ligase will join cohesive ends together wherever they are compatible. In many cases this will cause recircularization of the plasmid without any insert — a non-recombinant. This can be greatly reduced by pretreating the cut plasmid with alkaline phosphatase, which removes the phosphate groups from both ends. Linear polymeric molecules will also be formed, consisting either of several plasmids or foreign DNA molecules joined together or a mixture of the two. In a few cases one or more foreign DNA molecules will join the two ends of a plasmid, forming a circular recombinant. An equimolar

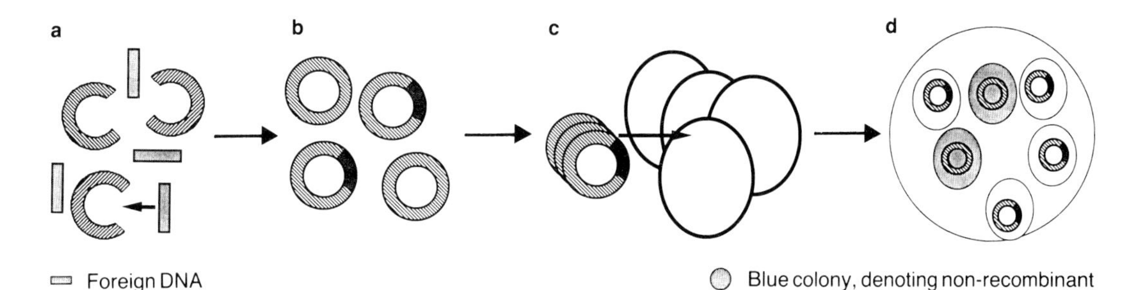

Fig. 10.2 Generalized scheme for cloning into plasmids. (a) The plasmid vector is cut by a restriction enzyme and mixed with the foreign DNA in the presence of the enzyme DNA ligase. (b) Some plasmids will simply recircularize but others will have the foreign DNA ligated into them. (c) The mixture of recombinant and non-recombinant plasmids is introduced into 'competent' host bacterial cells. (d) The bacterial cells are plated on to a medium containing antibiotic and colour selection reagents. Only those bacteria carrying the plasmid are antibiotic resistant and will grow. Recombinant plasmids have their colour selection gene interrupted and form white colonies. Non-recombinants form blue colonies.

ratio of plasmid to insert will produce the most recombinants. Once ligation has been performed the plasmids need to be introduced into host *E. coli*. This requires bacteria with leaky cell walls which will allow the plasmid DNA into the cell. Such 'competent' bacterial cells are made by incubating the bacteria in a solution of calcium ions. The ligation mixture and the competent bacterial cells are mixed together to allow the DNA to rest on the cell wall. The bacteria are then 'heat-shocked' for a few moments which allows the plasmid to enter the cell. The bacteria are then plated out on a selective medium and incubated at 37°C. Only bacteria which acquire a circular plasmid carrying an antibiotic resistance gene will be able to grow. Those carrying the recombinant plasmids can then be identified by colour. Recombinant colonies can be grown up in larger liquid cultures and the plasmids can be recovered from them. Digestion of the recovered plasmid with the original cloning restriction enzyme should release the insert and this can then be separated from the plasmid by electrophoresis.

Bacteriophage vectors

BACTERIOPHAGE LAMBDA (λ)

Viruses which infect and replicate within bacteria are known as bacteriophages or, more simply, phages. They consist of a core of DNA surrounded by a protein coat. The most commonly used phage vector is *E.coli* bacteriophage lambda (λ). Having infected the bacterial cell, λ phage DNA becomes circularized by pairing of the cohesive ends (the cos ends). The cos ends are DNA sequences at either end which join, either to circularize the phage or to cause phages to form large concatemers. The phage expresses the genes which cause replication of the DNA and synthesis of the packaging proteins. The cos ends are recognized by the packaging machinery, which combines the DNA genome of the phage with the protein envelope to form complete phage particles. The DNA and protein coats are assembled and the bacterial cell is lysed. The liberated phage particles are then able to transfect adjacent bacteria. This is termed the 'lytic' cycle. A small number of transfected bacterial cells will go into an alternative cycle, the 'lysogenic cycle'. In this case the phage DNA becomes incorporated into the host chromosomal DNA and does not multiply as in the lytic cycle. The choice between lysis and lysogeny depends upon an intricate balance between host and bacterial factors.

Bacteriophage λ was one of the earliest phage vectors to be used for cloning (Fig. 10.3). It has a genome size of about 50 kb, including a central section which is not essential for its lytic replication in *E. coli* but only functions to ensure that the phage is integrated into the host DNA during lysogeny. Special strains of λ have been created with *EcoRI* sites at either end of this central section. When the phage is digested by *EcoRI*, three fragments result: a left 'arm', a right 'arm' and

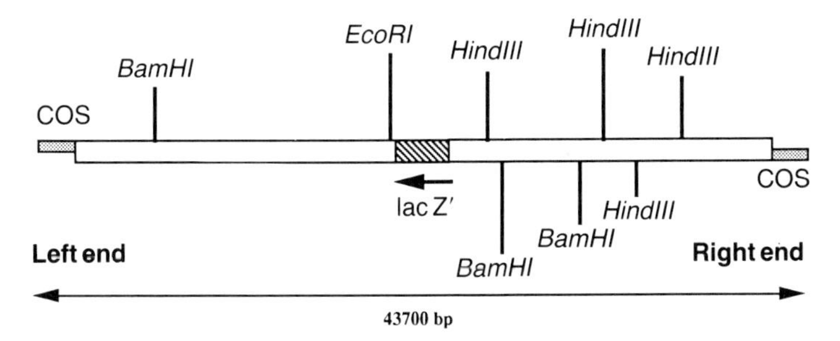

Fig. 10.3 The λ bacteriophage vector λgt11. Cloning into this vector is achieved by excising a central part (the stuffer fragment) and ligating in an insert that is of sufficient size to allow the recombinant phage to package. The phage is flanked by cos sites which are necessary for packaging and the lacZ′ gene which allows for colour selection of recombinants.

the small 'stuffer' fragment. The 'stuffer' is the central section, which is not essential for lytic replication. The 'arms', being much larger, are easily separated from the stuffer. Since λ phage will only package into its protein envelope if its DNA is about 50 kb in size, only recombinant molecules of a left and right arm joined by a foreign insert of the correct size will become viable.

Phage vectors have the advantage of being able to accommodate larger inserts than plasmid vectors — up to about 20 kb. In common with plasmids, a large number of different phage vectors have been engineered, all generally based upon λ. Again as with plasmids, they frequently contain a polylinker and some method of distinguishing recombinants from non-recombinants. This is often the same β-galactosidase system that has already been described. More recent phage vectors may also contain promoters for RNA polymerases, allowing their host bacteria to express the protein coded for by any insert. The same promoters may also be used for the *in vitro* synthesis of riboprobes — RNA probes for use in northern and Southern blotting and *in situ* hybridization. There are also phage vectors that contain an internal portion of DNA which, when excised, will then behave as a plasmid. Although it is easier to manipulate simultaneously large numbers of clones in phage vectors, growing single plasmids in bulk is much easier than growing single-phage clones.

Cloning foreign DNA into a phage vector follows a scheme similar to plasmid cloning (Fig. 10.4). The phage vector is cut with one or two restriction enzymes, and any stuffer fragments are separated from the arms. The foreign DNA is cut with the same enzymes and ligated into the arms, using a DNA ligase enzyme. This ligation reaction will also ligate the cohesive (cos) ends of the phage arms, forming large polymeric molecules. The phages are then 'packaged' into their protein coats, using specially prepared 'packaging extracts' prepared from transfected *E. coli*. Once packaged the phages are mixed with host cells to allow transfection. Phages are plated out together with untransfected cells on to an agar petri dish. The untransfected bacteria form a lawn on the

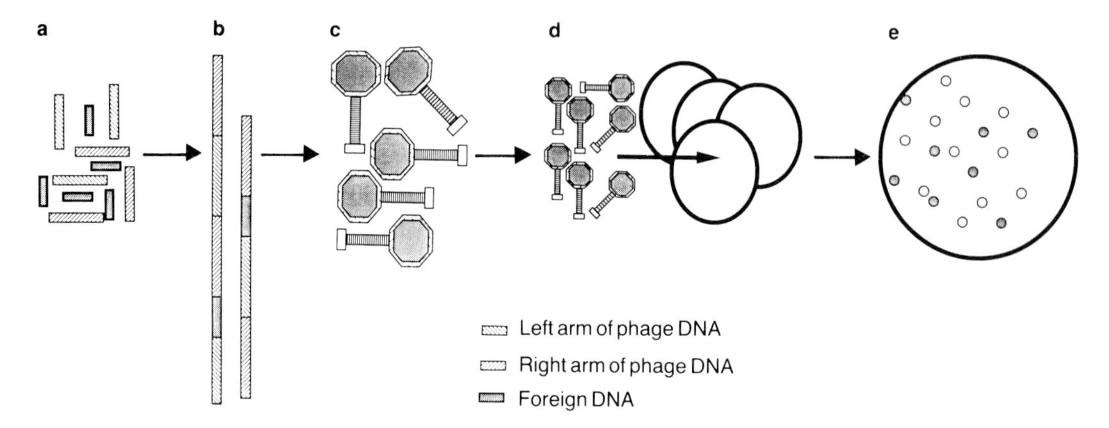

Fig. 10.4 Generalized scheme for cloning into λ bacteriophage. (a) Phage cut by a restriction enzyme forms left and right arms. These are mixed with the foreign DNA and a DNA ligase enzyme. (b) Large concatemers form consisting of recombinant and non-recombinant phage molecules joined end to end. (c) The *in vitro* packaging reaction cuts the concatemers into single phage molecules and assembles complete phage particles. (d) The phage particles are introduced into bacterial host cells and (e) plated out on to a medium containing colour selection reagents together with more bacteria not containing phage which form a lawn of bacteria on the plate.

agar. Infected cells lyse and the liberated phage are able to transfect adjacent cells. Gradually an area of lysis forms around the original single transfected bacterium. These 'plaques' contain high concentrations of a single-phage clone. With the β-galactosidase selection system, recombinant plaques are white whilst non-recombinant plaques are blue.

BACTERIOPHAGE M13

Bacteriophage M13 is a single-stranded filamentous phage which has a life cycle which has proved to be useful in sequencing DNA. Within the host bacterial cell it exists as a double-stranded DNA molecule similar to plasmid. Single-stranded phage particles are packaged and this single-stranded DNA may be used as a template in sequencing reactions. Foreign DNA can be cloned into the double-stranded M13 vector in the same way as into a plasmid vector, and competent bacterial cells can be transformed in a similar way and then plated out on to a lawn of untransfected bacteria. The plaques that form contain single-stranded M13, which can be isolated using standard bacteriophage isolation procedures. How these single-stranded molecules are used in sequencing DNA will be explained in Chpater 11.

Cosmid vectors

Cosmid vectors are an artificially engineered combination of phage and plasmid vectors. They include an antibiotic resistance marker, a cloning site which will accept foreign DNA of up to 50 kb in size, and the DNA sequences (cos sequences) needed for packaging the DNA into bacteriophage particles. Great care needs to be taken when preparing DNA for cloning into a cosmid to keep its molecular size high, so that, when it has been partially digested by restriction enzymes, its molecule size is within the range 35–50 kb. DNA for cosmid cloning is therefore prepared taking care to avoid the mechanical shearing that might reduce its molecular size. A series of trial restriction enzyme digests is performed to determine the ideal reaction time needed to produce a partial digest with resulting fragments in the range 35–50 kb. These fragments are overlapping so that the end of one cosmid clone is also represented at the beginning of another. As will be explained when discussing gene cloning (Chapter 11), an overlapping series of cosmid clones can be useful in creating a map of a specific area of genomic DNA. As with bacteriophage λ, ligation of the foreign DNA into a cosmid vector produces concatemers of vector and insert. The same packaging reagents can be used to package both cosmids and bacteriophage λ. The cos sites are recognized by the packaging reagents and the cosmid is packaged into a phage particle. These can then be transfected into host *E. coli* in the same way as with normal phage, and the cosmid DNA becomes circularized via the cohesive ends of the cos sites. Although cosmids could be introduced into host bacteria by transformation, as is done with plasmids, transfection is more efficient than transformation of large DNA molecules. The cosmid does not contain the genes needed for it to replicate phage particles but it is able to replicate as a plasmid. The size of the cosmid vector itself is 4–6 kb in length. This is too small for packaging so only cosmids which contain insert will be transfected into a cell. The cosmid will express its antibiotic resistance gene so that, if the host bacteria are plated on to agar containing the antibiotic, all of the colonies which grow should contain recombinant cosmids.

Yeast artificial chromosomes

Yeast artificial chromosomes (YACs) have been developed to allow the cloning of very large fragments of DNA, up to 1 megabase in size. These vectors make possible the cloning of a large piece of human DNA into a yeast DNA background. DNA fragments larger than the 50 kb limit of cosmids can be cloned into YACs. YACs are technically difficult to construct, but suitable clones can often be obtained that have already been constructed by YAC specialist laboratories. Like cosmids, YACs allow the mapping of larger regions of the genome than do conventional phage and plasmid clones. YACs are large enough to contain whole genes with all of their associated control elements. For example, the cystic fibrosis gene and all of its control regions can be accommodated within a single YAC, although the gene spans over 250 kb of DNA.

Constructing genomic libraries

A DNA library is a specific collection of DNA molecules cloned into a suitable vector with the aim that all of the DNA from the starting material is represented in clones several times over (Fig. 10.5). A phage total human genomic library, for example, is made by partially digesting total human DNA with a frequent-cutter restriction enzyme and selecting fragments of up to 15–20 kb. A phage vector is digested with the same enzyme and the human DNA is ligated into it. As well as total human genomic libraries, it is possible to construct libraries from cell lines containing a single, or part of a single, human chromosome. Using flow-sorting techniques, individual human chromosomes can be selected out. These can be introduced into an animal cell line, e.g. mouse. A library made from these cells will contain the mouse 'background' DNA with the only human DNA being from the selected chromosome. Libraries may also be made from the DNA of an individual with a chromosome deletion or insertion.

As discussed above, cosmid libraries may also be constructed to contain genomic DNA inserts of a larger molecular size. Care needs to be taken to maintain the molecular size during preparation of the DNA and digestion conditions are adjusted so that the enzyme used does not cut at every available site. This means that the DNA molecules cloned into a cosmid library form an overlapping series. If a single cosmid has been selected, perhaps because it contains an interesting region of the genome, a cosmid containing the DNA which is adjacent on the genome can be identified by screening the cosmid library with a probe made from one of the ends of the original cosmid (see Chapter 11).

COMPLEMENTARY DNA (cDNA) LIBRARIES

A cDNA library contains DNA molecules which are complementary to the mRNA population of the cell types from which it was originally made. A tissue or cell type will be chosen which expresses the genes being studied. Total RNA is isolated first and mRNA separated from it. A common feature of mRNA that most species have is a poly-adenine tail at the 3′ end. Chromatography columns can be prepared using poly-T (or oligo-dT) molecules covalently bound to Sephadex beads. When the total RNA is brought into contact with the column under high-salt conditions the poly-A^+ mRNA binds to the oligo-dT. The remaining poly-A^- RNA can then be washed off. The mRNA is then eluted from the column by reducing the salt concentration.

Messenger RNA itself is easily degraded by abundant RNases and, since it is single-stranded, cannot be cloned into a vector. To clone a representative RNA library, a DNA copy is therefore made of each of the RNA molecules in the population. The first strand of DNA is made using the enzyme reverse transcriptase (Fig. 10.6). This enzyme will

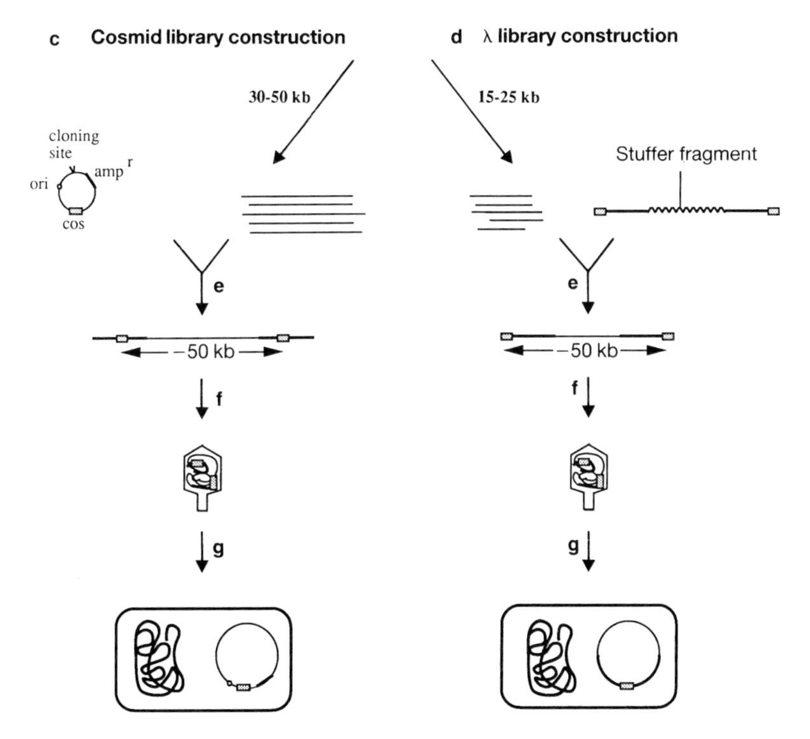

Fig. 10.5 Generalized schemes for cosmid and λ library construction. (a) Genomic DNA is partially digested with the selected restriction enzyme. (b) The DNA fragments are size fractionated by gradient centrifugation. (c) For cosmid library construction the cosmid vector is linearized at its cloning site. (d) For λ library construction the λ vector is digested and the arms separated from the stuffer fragment. (e) For both systems vector and insert (30–50 kb for cosmid library and 15–25 kb for λ library) are ligated together. (f) Concatemers of vector and inserts are packaged into λ particles. (g) DNA constructs are transfected into host *E. coli* for further propagation.

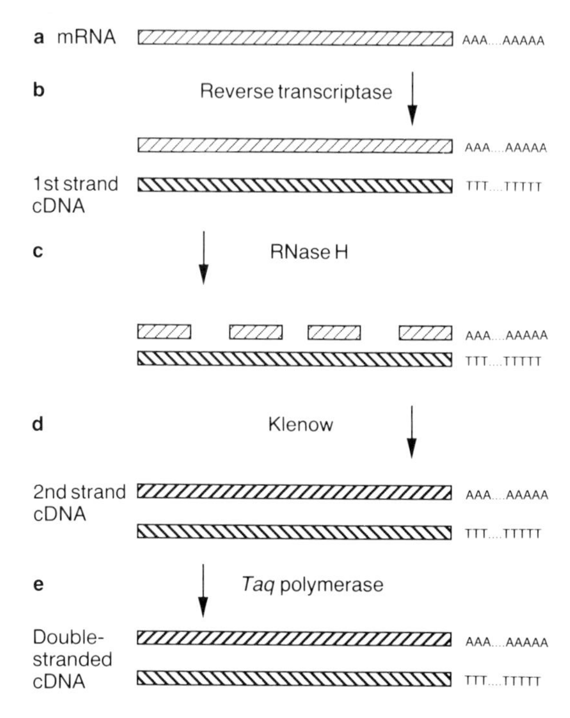

Fig. 10.6 Complementary DNA synthesis using oligo dT primers. (a) The oligo-dT primer anneals to the poly-A tail of the mRNA. (b) The first strand cDNA copy is then synthesized by reverse transcriptase. (c) RNase H is used to nick the RNA in the RNA-DNA hybrid. (d) Klenow then replaces the RNA using the islands of RNA remaining as primers. (e) *Taq* polymerase fills in any recesses left at the 3′ ends of both DNA strands.

make a copy of an RNA molecule by virtue of its 3′ to 5′ polymerase activity. The enzyme needs to be primed by annealing a small DNA molecule to the template RNA. This may be an oligo-dT molecule which will anneal to the 3′ poly-A tail of the RNA. First-strand synthesis then begins at the extreme 3′ end of the RNA molecules. Since the enzyme may fail to complete transcription all the way through to the 5′ end, this strategy may result in a library in which the 5′ ends of the mRNA population are under-represented. An alternative approach to this problem is to prime with random hexanucleotides. These will anneal at various sites along the mRNAs and increase the representation of 5′ regions but, since they cannot anneal to the poly-A tail, this is at the expense of representation of the 3′ end. It is also possible to use a combination of the two strategies to get the most complete representation.

Once the first DNA strand has been synthesized, the RNA is partially removed from the new DNA using RNase H. This leaves behind small islands of RNA which prime the action of a DNA polymerase to synthesize the second strand. Next, the double-stranded cDNA molecules are treated with a methylase which recognizes the same nucleic acid sequence as the endonuclease which will be used to clone the cDNA into the vector. This methylates the internal restriction sites within the cDNA, which will protect them when restriction digestion is used to form cohesive ends for cloning. Small double-stranded 'linkers' which contain the same recognition site are then ligated on to the end of the cDNAs. These are then digested with the restriction endonuclease to leave cohesive or 'sticky' ends (Fig. 10.7). Any identical recognition sequences within the cDNA should remain intact as a result of the methylation and should therefore be protected against endonuclease activity. The vector, usually a phage vector, is digested with the same restriction endonuclease, and the cDNA is ligated into it. The phage is then packaged into their protein envelopes and transfected into a suitable strain of host bacteria. The bacteria are then plated out on to a suitable medium and, after incubation, the presence of phage will be seen as plaques of lysis on the lawn of bacteria. Each plaque will represent where one single phage transfected one bacterium and then grew up to transfect those bacteria around it. Each plaque will contain a single 'clone' containing a single cDNA.

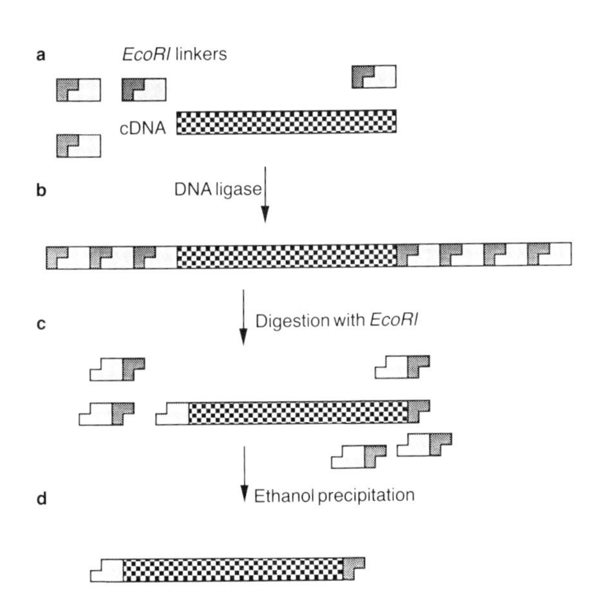

Fig. 10.7 Addition of synthetic linkers to cDNA to create restriction sites for cloning. (a) The cDNA is first treated with *EcoRI* methylase which methylates *EcoRI* recognition sequences within the molecule protecting them from later digestion. (b) Small DNA molecules containing recognition sites for the enzyme *EcoRI* are then ligated on to the ends of the cDNA molecules. This forms concatemers of cDNA and multiple linkers. (c) After digestion with the restriction enzyme *EcoRI* the sites within the linkers are cleaved leaving a cDNA with *EcoRI* sticky ends. *EcoRI* sites, if any, within the cDNA have been methylated and do not cleave. (d) Ethanol will precipitate only the large molecular weight cDNA molecules leaving the small linker fragments in solution. The cDNA molecules can then be cloned into a suitable vector.

Screening libraries to identify genes

Every clone in a cDNA library represents the mRNA from a gene which was expressed in the original tissue or cell. Genomic DNA libraries contain a large number of clones which originated from non-coding DNA; the number of clones which contain coding regions is therefore only a small proportion of the total. The usual approach, in cloning a gene about whose protein something is known, is to screen a cDNA library. Once the cDNA has been identified, then a total genomic library can be screened, using that cDNA to get information about the number of introns in the gene and about its 5′ controlling regions.

The protein information used in the first cloning experiments was that the protein itself is produced in huge quantities by certain specific cell types. Both globin and ovalbumin cDNA were identified because their cDNAs comprise as much as 50–90% of total mRNA in certain cells. For most cDNAs, representing lower-abundance mRNAs, some form of probe is needed to identify the relevant clone from a cDNA library. If a cDNA for the gene of interest from another species is available, it may have sufficient homology to the human gene to detect its cDNA in a library. Similarly, a human cDNA for one protein from a protein family may have regions of homology with the cDNAs for other members of that family and may be used to identify them from within a cDNA library.

SCREENING LIBRARIES USING OLIGONUCLEOTIDE PROBES

One of the most common strategies for cDNA library screening is to make a small oligonucleotide probe whose sequence is designed from knowledge of the protein amino acid sequence. From the peptide sequence data, the DNA sequence that codes for that part of the protein can be predicted. It is not necessary to know the sequence of the entire protein as the sequence of just a few of the fragments generated by digestion with trypsin may be sufficient. A region is chosen in which there is a run of amino acids for which there is minimal redundancy, i.e. where only one or two codons could code for each of the amino acids. Methionine, for example, is coded for by only one codon and phenylalanine by only two, whereas leucine and serine may be coded for by any of six codons each. It would be much more difficult to predict the correct nucleic acid sequence from a peptide sequence rich in leucine and serine than from a sequence rich in methionine and phenylalanine. By studying the coding sequences of those genes that have already been cloned, it has been found that certain codons are used more frequently than others to code for a particular amino acid. As more and more genes are being cloned, these codon frequency tables are being constantly updated. It has also been found that coding regions have a reduced frequency of GC pairs and an increased frequency of GT pairs compared with what would be expected by chance and what is seen in non-coding DNA. A 'best-guess' oligonucleotide probe can be made based on knowledge of the frequency of codon usage and the increased likelihood for GT pairs and reduced likelihood for GC pairs. Alternatively, a series of molecules can be synthesized which represent all possible alternatives such that somewhere in the mixture must be the correct sequence. The techniques used for making oligonucleotide probes are discussed in Chapter 9.

Once suitable oligonucleotide probes have been synthesized, the library is plated out on to agar plates (Fig. 10.8). The numbers of phage on the plate are adjusted so that the maximum number of plaques can be grown on each plate without plaques running together. Nylon hybridization filters are placed on to the plates. Some of the phage DNA from each plaque transfers to the filters, forming a replica of the plaque pattern on the filter. The DNA on each filter is then alkali-denatured to make it single-stranded and is neutralized, and the filters are baked to irreversibly bind the DNA to the filter. The synthetic oligonucleotide probe is radiolabelled by transferring a radiolabelled phosphate on to the 5′ end, using the enzyme DNA kinase (see Chapter 7), and hybridized to the replica filters. Plaques which hybridize to the probe are picked from the original plate. To be sure that only the correct plaques are picked, these are then plated out again but with fewer plaques on each plate to allow secondary screening. After secondary and sometimes tertiary screening or more, a number of clones may be identified which will always hybridize strongly to the original probe. DNA is prepared from larger-scale cultures to allow the clone to be further characterized.

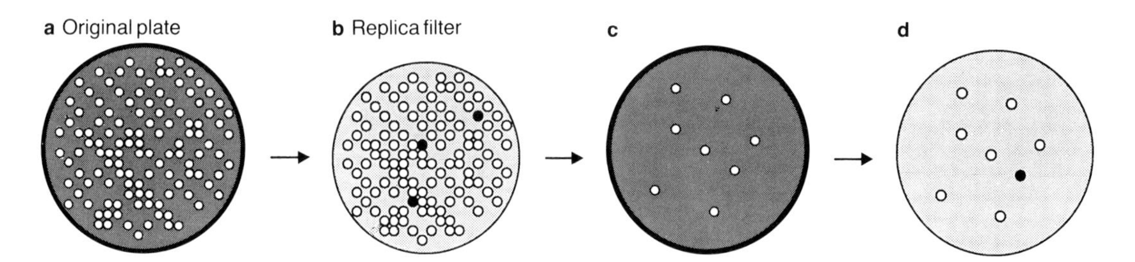

Fig. 10.8 Generalized scheme for cDNA library screening. (a) The library is initially plated out at high density. Each plaque (○) represents a single clone. All of the phage within a single plaque contain clones of the same DNA molecule. (b) A nylon filter is placed on to the plate; some of the phage DNA from each plaque transfers forming a replica on the filter. The filter is hybridized to the probe designed to detect the cDNA of interest. The autoradiograph of the filter is matched to the original plate and the positive plaques (●) are picked. (c) At this stage it is not possible to pick individual single plaques so the positives are plated out again at lower density. Once one or more positive clones have been singled out the phage can be grown in bulk to allow further analysis of the clone. (d) After a second round of screening using the same technique, individual plaques can be picked from the secondary plate. If plaques on the second plate are too close together to pick single plaques clearly, a third round of screening may be performed.

SCREENING cDNA LIBRARIES USING ANTIBODIES

If the cDNA has been cloned into an 'expression vector', the protein encoded by each cDNA may be produced by the host bacterium. The library can then be screened in a way similar to that using oligonucleotide probes but, instead of using DNA probes, a labelled antibody directed against the protein in question is used. There are a number of pitfalls with this approach. If the cDNA in question is not a full-length copy of the original mRNA, then an incomplete protein may be synthesized which will not be recognized by the antibodies. Similarly, if, *in vitro*, the protein does not fold in the way it does *in vivo*, it may not be recognized by the antibodies. These problems may be overcome by using polyclonal rather than monoclonal antibodies. Sometimes the protein of interest interferes with the biochemistry of the host bacterium, which then fails to grow normally.

FURTHER CHARACTERIZATION OF A cDNA CLONE

Once a cDNA clone has been identified as a candidate for the coding sequence for a particular protein, further proof is needed of the identity of the clone. Since the clone has been selected from a cDNA library, it is almost certain to code for a gene. This could be confirmed by

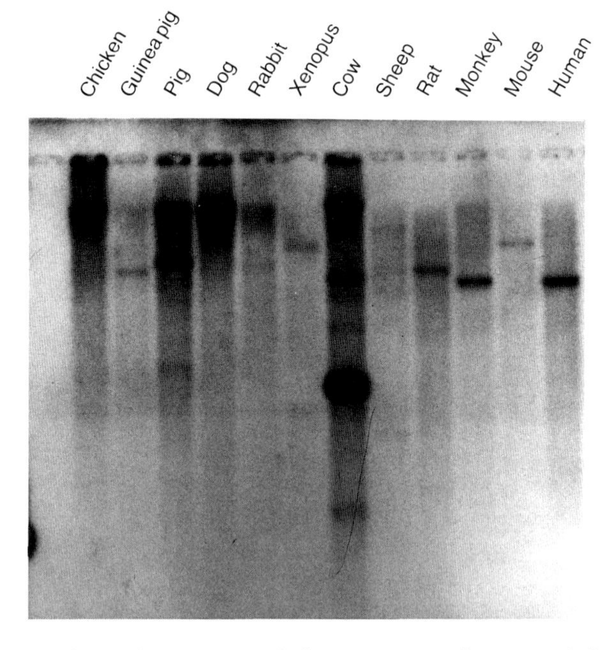

Fig. 10.9 Autoradiographic exposure of a human DNA probe (pCS.7) hybridized to a Southern blot containing DNA from a number of different species ('zooblot'). Cross-hybridization to DNA from different species is a good indicator of the presence of a gene, as coding regions are often conserved across species. In this case pCS.7 was found to contain the 5′ exons of the human proto-oncogene INT1L1.

hybridization to a Southern blot containing DNA from a wide range of species — a 'zooblot' (Fig. 10.9) (or, if the zooblot contains male and female pairs of DNA, a 'Noah's ark' blot). Genes are frequently highly conserved between species and so homologous sequences detected on a zooblot are evidence that the original clone contains a coding region. If the cDNA does not appear to be full-length, i.e. if it does not represent

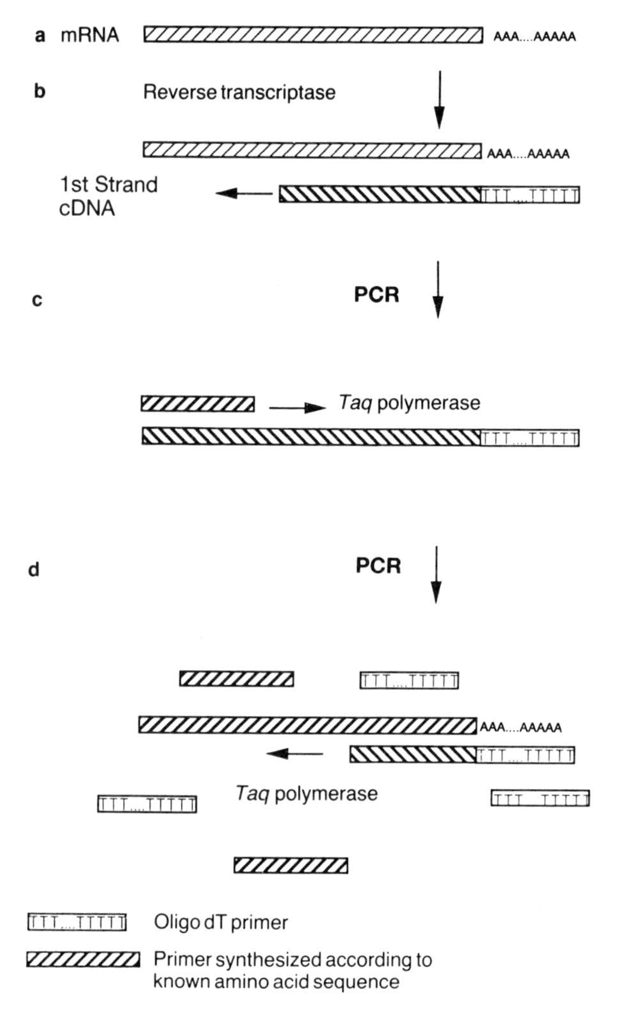

Fig. 10.10 General scheme for using the polymerase chain reaction (PCR) to aid the cloning of specific cDNAs. (a) In a preliminary reaction oligo-dT primers anneal to the poly-A tail of the mRNA. (b) The first strand cDNA copy is then synthesized by reverse transcriptase. (c) PCR is then performed using oligo-dT primers and primers designed to match the 5′ end of the cDNA to be amplified. Some information about the amino terminus of the protein is needed to allow this primer to be designed. (d) In subsequent rounds of amplification the oligo-dT primer will bind to the poly-A tail of the sense strand of the new DNA molecules whilst the cDNA specific primer will bind to the 3′ end of the antisense strand. After sufficient rounds of amplification the cDNA PCR product can be cloned into a suitable vector and analysed.

the full length of the mRNA, it can be used as a probe itself to rescreen the cDNA library. This may produce clones which contain the missing portions. The cDNA can be hybridized to Northern blots or used in *in situ* hybridization experiments to show that the mRNA is expressed in those cell types which produce the original protein. The cDNA may be translated and transcribed *in vitro* to analyse the protein product or to show that it codes for the original protein, or hybridized to the native mRNA in an *in vitro* preparation to inhibit its translation – hybrid-arrested translation. Determining the nucleic acid sequence of the clone may show that regions of DNA sequence are colinear with regions of known amino acid sequence.

If more information is needed about the gene itself, e.g. the number and size of introns or the non-coding controlling sequences around the gene, the cDNA can be used to screen a genomic library. This should produce genomic clones which contain both the coding and non-coding sequences of the gene. Various techniques may then be used to compare the cDNA and genomic sequences.

Using the polymerase chain reaction to clone genes

The polymerase chain reaction may be used to clone a cDNA without the need to construct a cDNA library if some of the amino terminal sequence of the protein is known (Fig. 10.10). Messenger RNA is isolated from a cell type which expresses the protein. First-strand cDNA synthesis is then performed, using an oligo-dT primer which will anneal to the 3′ end of the mRNA and prime reverse transcriptase. Second-strand synthesis is then primed by an oligonucleotide designed from the amino terminal sequence, which will therefore anneal to the 5′ of the cDNA first strand. Polymerase chain reaction is then performed, using the 5′ oligonucleotide and the 3′ oligo-dT as primers. Since the 5′ oligonucleotide should be specific for the protein of interest, only the cDNA coding for it should be amplified.

11: Identification of genes whose protein product is not known

Gene mapping

It is possible to map the chromosomal location of a gene whose protein product is not known. For example, the genes for Duchenne muscular dystrophy and cystic fibrosis were first mapped to the X chromosome and chromosome 7 respectively by studies of chromosome rearrangements and linkage analysis before other techniques allowed the genes themselves to be found. Even when a gene has not been identified, knowledge of its chromosomal location may enable a mutated gene to be tracked through families. This may be used for prenatal diagnosis, identification of carrier status or, in the case of diseases which may manifest later in life, such as Huntington disease, identification of affected individuals.

Clues to gene location from chromosome abnormalities

The first diseases to be mapped to a specific chromosome were those on the X chromosome. Long before chromosomes had been discovered it was known that certain diseases, such as haemophilia, were inherited through the female line and affected only the male offspring. Females are usually heterozygous for the relevant gene and are unaffected. Occasionally, if the gene mutation is sufficiently common, e.g. the glucose-6-phosphate dehydrogenase deficiency found in certain African populations, there may be homozygous females who are affected.

Usually chromosomal abnormalities involve large portions of the genome, affect many genes and produce complex syndromes. Identification of the specific genes involved is impossible. For example, Down syndrome (trisomy 21) presumably arises due to overexpression of genes present on chromosome 21. Whilst overexpression of genes from chromosome 21 has been seen in Down syndrome, e.g. the gene for superoxide dismutase, it is not at present possible to say how this overexpression causes the phenotype. Presenile dementia occurs more commonly in Down syndrome than in the normal population; therefore, work to identify the gene or genes responsible is concentrated on chromosome 21.

Chromosome deletions and translocations often give a clue to the location of specific disease-associated genes. Small visible chromosome deletions are occasionally associated with syndromes that are usually inherited as single-gene defects. The Prader–Willi syndrome is a single autosomal dominant disease which causes severe mental retardation from early childhood. It is occasionally seen in patients who have a visible deletion near to the centromere on the long arm of chromosome 15. Retinoblastoma, a tumour of the retina, may be inherited in an autosomal recessive fashion or may appear to be sporadic. In some cases, both inherited and sporadic, it is associated with a visible deletion of the long arm of chromosome 13.

Unbalanced translocations result in the inheritance of significant amounts of additional genetic material and, if not lethal, may cause multisystem syndromes. Balanced translocations are often asymptomatic, since the total amount of genetic material is unchanged. There is, however, usually some loss of genetic material at the break point. This usually occurs in non-coding DNA and has no physical effect, but occasionally a balanced translocation is associated with a deletion of a gene which happens to be at the break point. The rare DiGeorge syndrome is inherited as an autosomal dominant disease but has been described in individuals having balanced translocations and deletions involving chromosome 22. Work to identify the gene or genes responsible is therefore initially concentrating on this chromosome.

Duchenne muscular dystrophy, an X-linked disease usually only seen in males, has been described in females with translocations from autosomes to the X chromosome. X inactivation, the mechanism by which only one X chromosome is active in the normal female, is usually random, with approximately half of the cells expressing each of the X chromosomes. When an X chromosome carries additional autosomal material, the other, normal, X chromosome appears to be preferentially inactivated. This is probably because cells in which the X chromosome bearing autosomal material is inactivated are not viable and are selected against during growth and development. Duchenne muscular dystrophy in females is associated with insertion of genetic material on to the short arm of the X chromosome, which suggested that the gene would be found at Xp21. The gene was later identified within this region.

Linkage analysis

Linkage analysis is a technique used two map genes whose protein product is unknown. Two genetic loci are said to be linked if they segregate in any population more frequently than would be expected by chance. Consider two genes A and B, each with two alleles, 1 and 2. If these genes are on different chromosomes, there will be no relationship between inheritance of a particular allele of A with a particular allele of B. This also applies if the genes are far apart on the same chromosome, since the phenomenon of genetic recombination, or crossing-over, at meiosis, means that they will be inherited in the same way as if on different chromosomes. If the two genes are on the same chromosome but very close together, such that recombination virtually never occurs between them, then inheritance of a particular allele of gene A will always be associated with inheritance of a particular allele of gene B. These two genes would then be said to be linked.

Examples of linkage between genes which can easily be distinguished by their phenotype are rare. One long-recognized example is the nail–patella syndrome. This syndrome has an autosomal dominant inheritance and causes dysgenesis of the nails and dysgenesis or absence of the patellae. The gene responsible is linked to the ABO blood group locus. In any affected family, inheritance of the nail–patella gene is associated with inheritance of a particular ABO antigen. This could be used to predict risk for offspring by bloodtyping parents or for prenatal diagnosis by blood-typing the fetus, from either chorionic villi or blood. In practice, the risks of prenatal diagnosis are not justified since the disease is relatively benign.

Linkage between known structural genes is at present so rare that linkage analysis must be undertaken using other genetic markers. The most commonly used are restriction fragment length polymorphisms (RFLPs), which can be detected using probes for anonymous stretches of DNA whose chromosomal location is known. If genomic DNA is digested with a restriction enzyme, Southern blotted and hybridized to a radiolabelled probe, that probe will detect a fragment whose size depends upon the distance between the two restriction sites either side of it. A mutation may create or destroy a restriction site but, if this site is in non-coding DNA, will have no other effect on the individual. The distance between the two restriction sites on either side of the region detected by the probe may be different on each of the two chromosomes (Fig. 11.1). If an individual is heterozygous for this restriction length polymorphism, then the probe will detect two different fragment lengths on the Southern blot, one derived from each chromosome. The region of the chromosome detected by the probe can be determined by *in situ* hybridization to metaphase spreads or by hybridization to DNA from cell lines containing chromosome deletions. These RFLPs obey the same Mendelian laws of inheritance as do structural genes (see Fig. 8.6). The large number of restriction enzymes available means that RFLPs are very common throughout the genome. The two-band polymorphism described is the simplest case. Often a probe will detect multiple bands. These may arise because the restriction sites occur within the region of DNA detected by the probe as well as either side of it. In the case of large probes, multiple restriction sites may mean that the probe detects several bands, some of which may be polymorphic whilst others are non-polymorphic, constant bands (Figs 11.2 & 11.3). The areas of homology between the X and Y chromosomes sometimes cause probes derived from the X chromosome to detect Y-specific bands. Nevertheless, provided that a large enough pedigree is available for study, the inheritance of bands can be followed through several

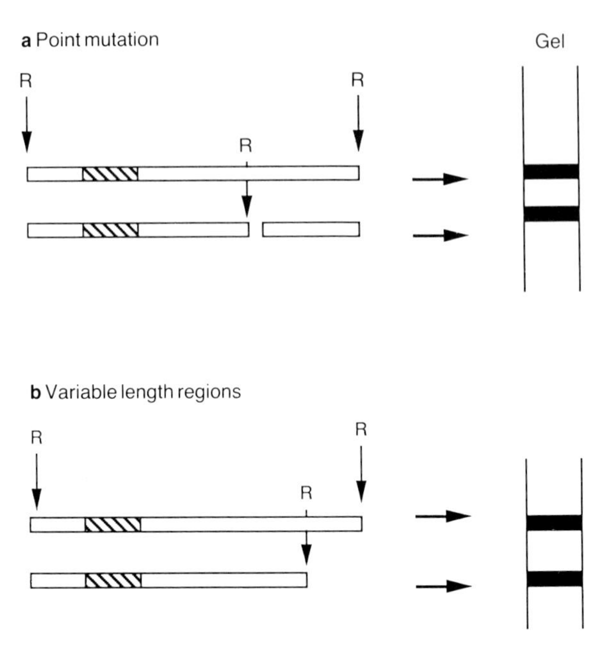

Fig. 11.1 Formation of restriction fragment length polymorphisms (RFLPs). (a) A point mutation creates a new recognition site for the restriction enzyme. The probe detects a shorter restriction fragment when the substitution is present. (b) Deletion or insertion of a region of genomic DNA changes the distance between two restriction sites.

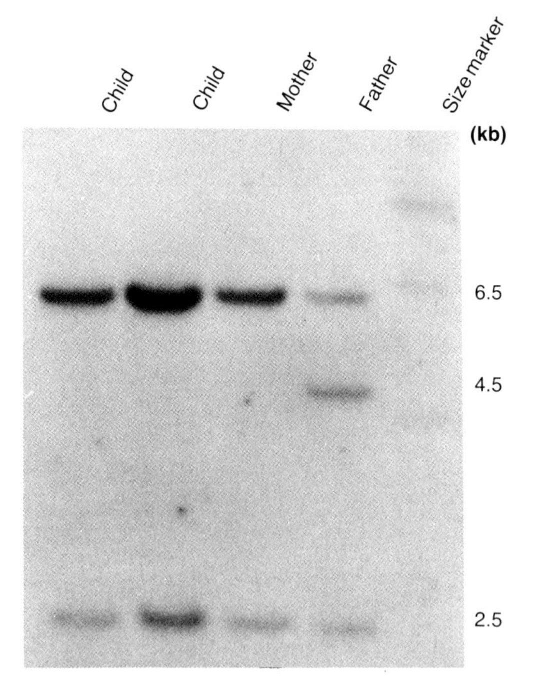

Fig. 11.2 Autoradiograph showing a *TaqI* polymorphism detected by the probe Met D. The polymorphic bands are 6.5 and 4.5 kb but the probe also detects an invariant constant band in all individuals.

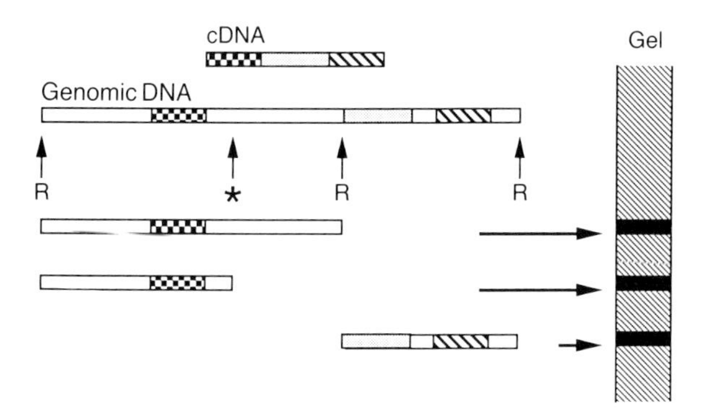

Fig. 11.3 In this case the cDNA probe spans a region which includes two restriction sites, one of which is polymorphic. In addition to detecting polymorphic fragments, the cDNA also detects a fragment which is constant in all individuals.

generations, and particular band patterns can be assigned to particular alleles and therefore to one of any pair of chromosomes.

Linkage analysis aims to determine how frequently a particular phenotype, usually a disease, is inherited with a particular RFLP allele. In the case of X-linked diseases, only probes present on the X chromosome need to be studied. If chromosome deletions or translocations give clues as to the location of the gene, then RFLPs from that chromosome will be studied first. If an RFLP and the gene locus segregate randomly in the pedigree under study, then they are probably on different chromosomes or far apart on the same chromosome so that recombination often occurs between them. If the RFLP is close to the gene locus, a particular allele will segregate with the disease more frequently, and, if they are at the same locus, they will always segregate together.

Statistical analysis of linkage studies

In the same way that a clinical trial requires a minimum number of subjects and controls for the results to reach statistical significance, a linkage study requires a pedigree of adequate size. The key feature is the number of informative meioses. For any individual mating to be informative for a particular RFLP, at least one of the parents needs to be heterozygous, or, in the case of X-linked diseases, the mother needs to be heterozygous. Statistical analysis of a linkage study for a particular RFLP can in some cases be done 'by hand', but in general the mathematics involved are sufficiently complex for one of a number of specifically designed computer programs to be used. Before the study is begun, it is usual to do a computer model linkage analysis with a theoretical, highly polymorphic probe and the relevant pedigree. Only if this shows that the number of meioses in the pedigree is adequate for linkage to be demonstrated in theory is it wise to embark on the study itself. Although some researchers have been lucky enough to demonstrate linkage to a disease by one of the earliest RFLPs studied, a linkage study is usually a lengthy and labour-intensive process.

LOGARITHM OF THE ODDS SCORE

The results of a clinical trial are expressed as the statistical probability that one treatment is better than another, with certain probability values (p) being generally accepted as showing a true difference. Similarly the results of a linkage study with a particular probe and RFLP are expressed as the likelihood that the RFLP is unlinked to, linked to or lies at a particular distance from a gene. These are calculated as odds scores, and these odds are usually expressed as a logarithm to base 10. An LOD score of +3 means that the odds in favour of linkage are 1000 to 1 or that there is only a 1 in 1000 chance that this is not true linkage and is a random association. An LOD score of −2 means that the odds are 100 to 1.

In the discussion above, the LOD score shows the likelihood that linkage is genuine and not just a random association. For a disease gene and an RFLP to be tightly linked, there will be no recombination between the two. Statistical analysis, in this case, answers the question, 'What are the odds that the data I have from my pedigree support linkage, with no recombination between the RFLP and the gene?' Of course, the ideal result of a linkage study is to find an RFLP which shows no recombination with the disease. However, it is also possible, when there are recombination events between the two loci, to use LOD scores to assess the likelihood that the frequency of recombination observed in the study is a genuine reflection of the frequency of recombination in the general population. The frequency of recombination between two loci is termed the recombination fraction. The recombination fraction is directly related to the physical distance between the two loci. The further apart the two are, the greater the chance of recombination between them. *Genetic distance* is expressed in centimorgans (cM) and is calculated from recombination fractions. One centimorgan is a genetic distance over which recombination occurs in 1% of meioses, a recombination fraction of 0.01. *Physical distance* is expressed in base pairs (bp). As a generalization, 1 cM is equivalent to 1 million bp but there are large variations within the genome. Chromosomes vary considerably in size. Physically larger chromosomes have more recombination events than smaller chromosomes. There are chromosomal areas in which cross-over events occur more commonly; these are known as 'recombination hot spots'. Since several recombination events may occur in only a short physical distance at a recombination hot spot, genetic distance here overestimates physical distance. Very little recombination occurs near to the centromere of any chromosome, so here genetic distance will underestimate physical distance. In some cases there are also differences in the recombination fraction between meioses in oocytes and in spermatozoa.

LOD scores, therefore, can also be used to answer the question, 'What is the likelihood that the recombination fraction between the two loci is θ?' where θ is a predicted recombination fraction. LOD scores can be calculated for various values of θ from 0, meaning no recombination at all between the RFLP and the gene and therefore linkage, through to 0.50, meaning independent assortment between the gene and the RFLP. Values between 0 and 0.50 allow the likelihood that the RFLP lies a certain genetic distance from the gene to be calculated. A recombination fraction of 0.01 means that recombination occurs between the RFLP and the gene in 1% of meioses. The value of θ which carries the highest LOD score is the most likely genuine recombination fraction.

An LOD score of zero means that linkage and no linkage are equally likely. A positive LOD score increases the likelihood whilst a negative LOD score decreases it. The significance values are universally agreed to be +3 to demonstrate linkage and −2 to demonstrate non-linkage. An LOD score of +3 gives a 95% overall probability. A higher likelihood is required for linkage to be accepted because the prior probability that two loci are linked is very low. To be convinced that two loci are linked, convention requires stronger evidence than to be convinced that they are not linked, since, in general, loci are not linked.

Once LOD scores have been calculated for a particular gene and RFLP at various recombination fractions, they can be plotted graphically, as shown in Fig. 11.4. Curve (a) has its peak LOD above +3 at a recombination fraction of 0. This is evidence for tight linkage and suggests that the gene and the RFLP are very close together on the chromosome. Curve (b) also has a peak LOD above +3 but this is at a recombination fraction of 0.2. This suggests that the RFLP lies at distance from the gene such that recombination will occur in 20% of meioses. This represents a genetic distance of 20 cM. As discussed above the

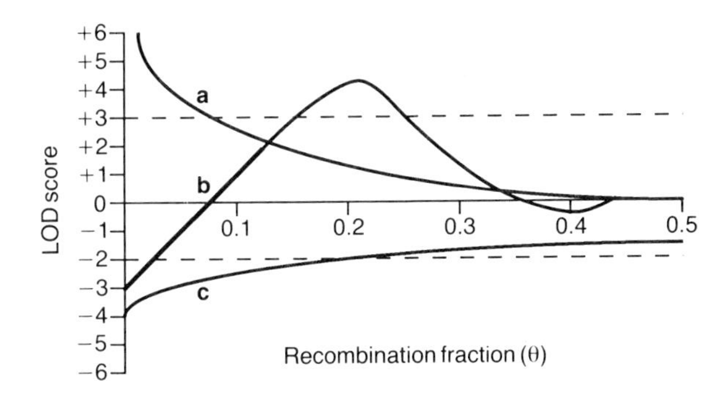

Fig. 11.4 Graphic representation of LOD scores calculated at various recombination fractions (θ). Curve (a) has its peak LOD above 3 at a θ of 0 indicating close linkage. Curve (b) has a peak LOD above 3 when θ is 0.2, suggesting a genetic distance of 20 cM between the RFLP and the gene. Curve (c) has a LOD of −2 or less when θ is between 0 and 0.2, excluding linkage closer than 20 cM.

physical distance may be about 20 million bp but will vary depending upon the size of the chromosome and the location on it. Curve (c) has an LOD of −2 or less at recombination fractions between 0 and 0.2, which excludes linkage closer than 20 cM.

LOD scores generated using the same RFLP in different pedigrees can be added together. Care needs to be taken, however, to ensure that the phenotype is identical in the two families. In certain cases, e.g. cystic fibrosis, there is little difficulty in diagnosis. In other cases, e.g. psychological disorders, diagnosis may be difficult. It also needs to be remembered that the same phenotype may arise in different pedigrees by different genetic mechanisms. Adding together LOD scores in these cases may actually reduce evidence of linkage when linkage does exist in one of the two groups.

MULTIPOINT LINKAGE ANALYSIS

So far we have considered linkage analysis with a single RFLP, but it is possible to analyse linkage data from several loci (several RFLPs). This is useful once the rough localization of the gene is known since it allows the gene locus to be mapped with respect to other RFLP loci. Once flanking markers have been established, the area between them can be studied in more detail to find the gene itself.

Informativity

A probe may detect an RFLP in the general population but, if every member of a particular pedigree under study is homozygous for one of the alleles, it will be impossible to use that RFLP for linkage analysis. Frequently a particular RFLP is not sufficiently informative in a pedigree to be able to determine whether or not it is linked to a disease locus. It is possible that the probe concerned may detect a more informative RFLP with another restriction enzyme. Other RFLPs may be found by hybridizing the probe to Southern blots of genomic DNA from random members of the population; this DNA should have been digested with other restriction enzymes. These Southern blots are often referred to as polymorphism searching blots or, more simply, polyblots.

Polymorphic repeat regions provide another system of polymorphisms which may be useful in linkage analysis. These are stretches of DNA, usually within non-coding regions, which are composed of a variable number of repeating elements. A common repeating element is the dinucleotide AC, where stretches of 20 or 30 base pairs are ACACACAC..., with complementary TGTGTGTG... on the complementary strand. The variability of the number of AC repeats is inherited in a Mendelian fashion. The presence of AC repeats in the area of interest can be detected by hybridization of a small TG or AC repeat oligonucleotide. The sequence of that probe can then be found and primers for PCR can be made to flank the AC repeat region (Fig. 11.5).

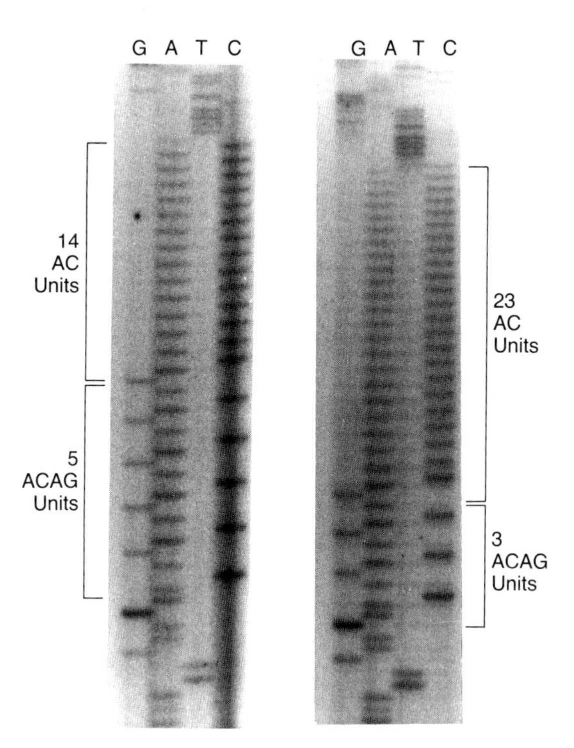

Fig. 11.5 Direct PCR sequence showing AC repeats of different lengths in two different individuals. Note that in this case there is also a second type of repeat sequence which is also variable in copy number in different individuals.

The polymorphism will be demonstrated by PCR amplification of that region producing fragments of variable length depending upon the number of AC repeats (Fig. 11.6).

It is also possible to sequence short stretches of DNA in the region of interest in the hope of finding a polymorphic point mutation which

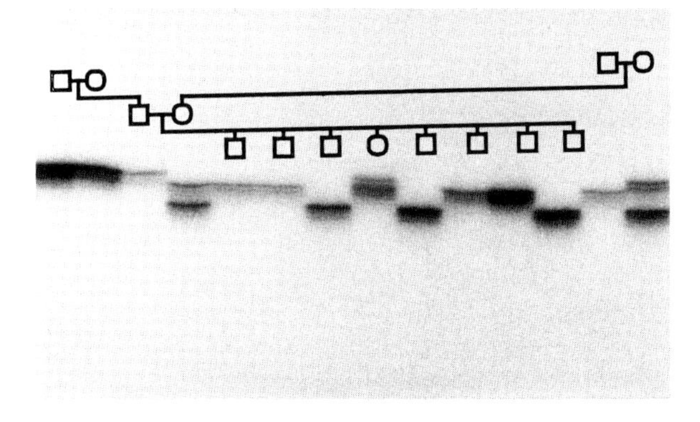

Fig. 11.6 Linkage studies using AC repeat analysis. The AC repeat sequence is amplified in different family members using PCR. The products are radiolabelled and then fractionated on a 6% denaturing polyacrylamide gel. The different sizes of the alleles are due to different numbers of AC dinucleotides in the repeat sequence.

does not create or destroy a restriction site. In the past, DNA sequencing was an expensive and time-consuming technique but new PCR technology has made sequencing of short stretches of DNA more efficient.

Moving from linkage to identification of the gene itself — chromosome walking and jumping

With luck a linkage study may identify one or more RFLPs which show no recombination with the disease locus. Mapping of the positions of linked markers relative to each other can be achieved with the aid of pulsed-field gel electrophoresis. By digesting genomic DNA with various rare-cutter enzymes and hybridizing the linked probes to the resulting large fragments separated by pulsed-field electrophoresis, an idea can be obtained of the relative positions of each of the markers to the others. Ideally an RFLP will be found close on either side of the disease gene itself. Even in this case, there may be quite some considerable distance between the gene being hunted and the RFLPs. For example, if the distance between them were 1 million bp, a genetic distance of 1 cM, it would require about 100 informative meioses to demonstrate one recombination event. Although this is very close in genetic terms, it is still much too large a region to look for genes by sequencing all of the intervening DNA.

There are a number of techniques for moving in to find the gene itself. An overlapping genomic library may be screened with one of the linked probes. Once the clone containing that region of DNA has been found, another probe can be made from the extreme end of it. If this is used to rescreen the library, an overlapping clone may be found moving in the direction of the gene being sought. Again, a new probe can be made from the end of this clone and the next overlapping clone found. This is termed 'chromosome walking'. It may be possible to assemble a collection of overlapping clones which span the region of interest, forming a so-called 'contig'. Chromosome walking may be done using an overlapping phage library but it is more commonly done using a cosmid library since the larger size of cosmids make the rate of 'walking' faster (Fig. 11.7).

In order to move directionally across the genome more rapidly, a technique related to chromosome walking called 'chromosome jumping' has been developed. Although there are several adaptations to this technique, a common method makes use of rare-cutter restriction sites, which are found throughout the genome at intervals of several hundred kb. Initially, specific jumping and linking libraries have to be constructed (Fig. 11.8). A jumping library is constructed by digesting the genomic DNA of interest with a selected rare-cutter enzyme to give large DNA fragments. These fragments are then circularized in the presence of a plasmid that has ends compatible to the rare-cutter enzyme. The plasmid contains a selectable marker. When it is incorporated into one of the large circular fragments, this acts as a tag. These fragments are then

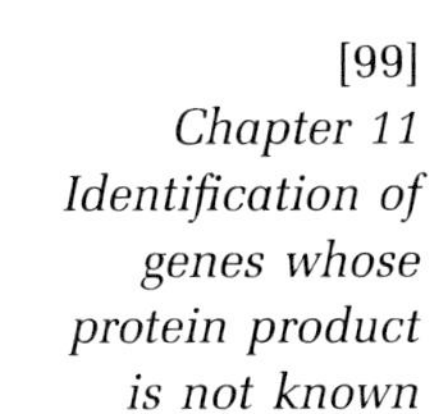

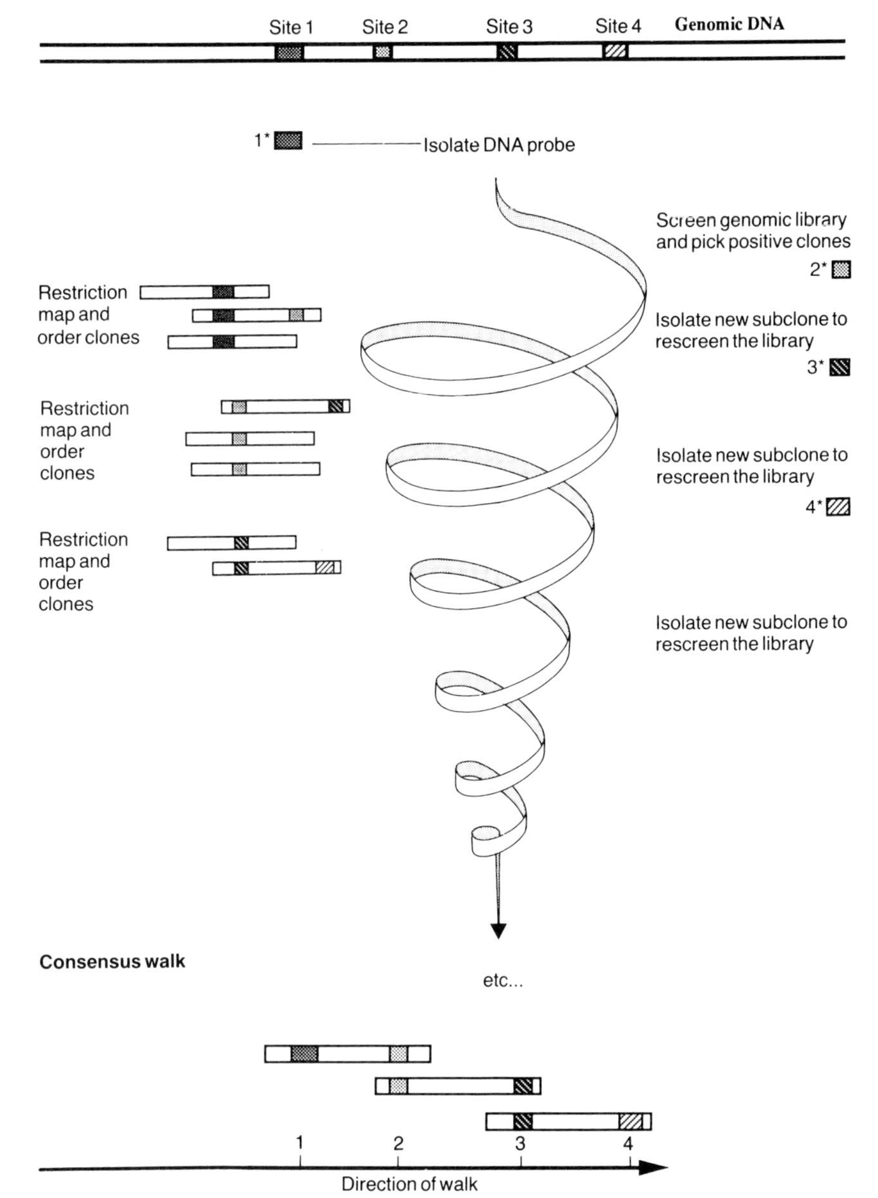

Fig. 11.7 Chromosome walking. Walking can be carried out with either bacteriophage λ or cosmids. Cosmids have become the first choice as longer walks can be made more rapidly.

digested with a frequent-cutter enzyme that does not cut within the plasmid tag, giving small linear fragments, some of which will contain the plasmid tag and be flanked by the rare-cutter enzyme and fragments. All of the fragments generated by this process are then cloned into a suitable λ phage vector. Only the clones containing the plasmid tag are

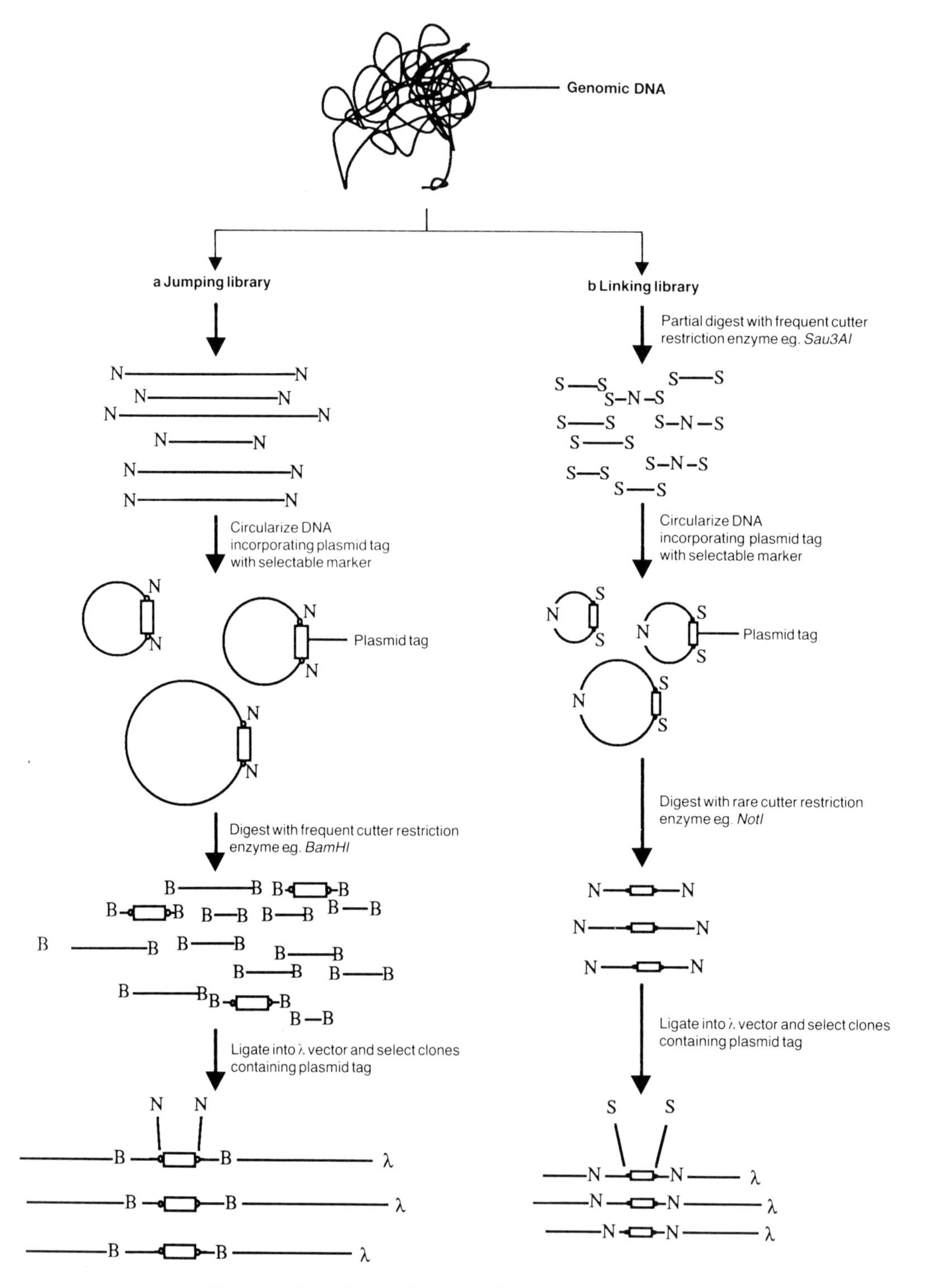

Fig. 11.8 Generalized scheme for the construction of (a) jumping and (b) linking libraries.

able to grow, giving a library of clones that contain a pair of rare-cutter sites that were originally several hundred kb part.

A second library is then constructed that contains the DNA-flanking rare-cutter restriction sites. This is called a linking library. Successive jumps over large distances are then made by screening alternately between both libraries. The starting-point is a clone that contains a rare-cutter site in a known position. This is used to screen the jumping library. A positive clone will contain the same rare-cutter restriction site plus a second one from some distance further away. This represents a large jump without the need to clone any of the intervening DNA.

To move further along the genome, this second site, with adjacent DNA, is subcloned and used to screen the linking library. From this clone, DNA from the other side of the rare-cutter site is subcloned and then used to screen the jumping library again. The net result is to move from rare-cutter site to rare-cutter site along the genome (Fig. 11.9). Confirmation of the results of each jump is usually obtained by restriction analysis using pulsed-field gel electrophoresis with the same rare-cutter restriction enzyme.

Identification of the gene itself might be achieved by looking for HTF islands. Many genes have GC-rich regions, especially at their 5′ ends. These contain large numbers of sites for the restriction enzyme *HpaII* (Fig. 11.10). This enzyme is usually a rare-cutter and generates large DNA fragments, but at these GC-rich regions it will generate tiny fragments (hence *HpaII* tiny fragment (HTF) islands). Another approach is to use candidate cosmids to screen cDNA libraries made from tissues expected to express the gene being sought. So, for example, in searching for a gene whose mutation causes muscular dystrophy, a cDNA library

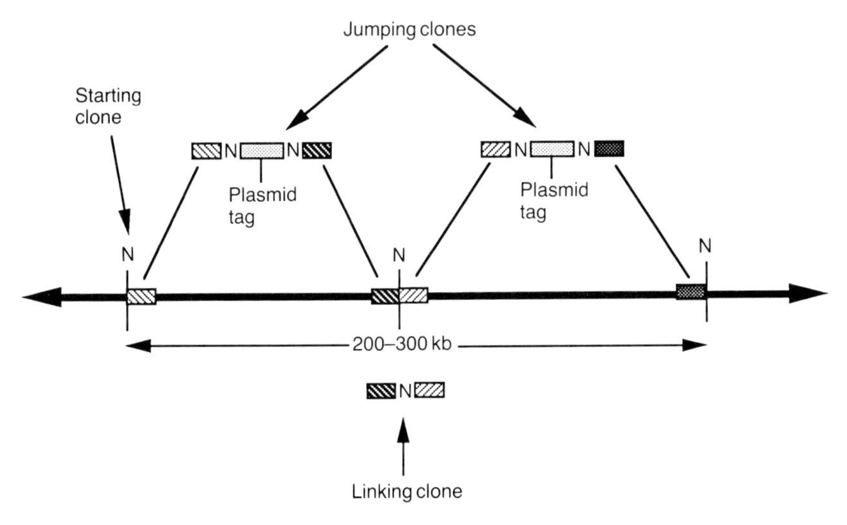

Fig. 11.9 The net result of two jumps. Note that the linking library is screened after each jump to provide the probe for the next jump.

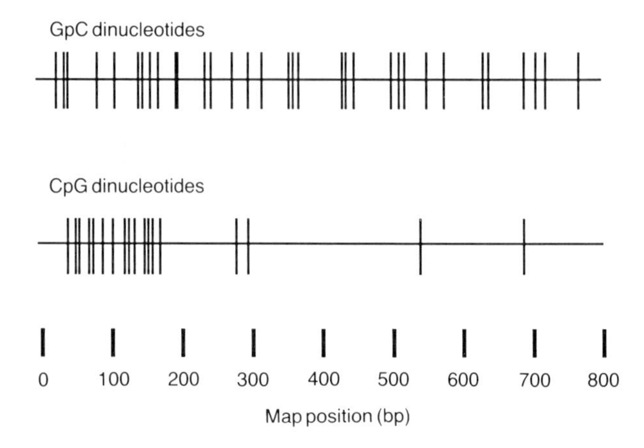

Fig. 11.10 CpG and GpC distribution in a gene sequence. Each vertical line represents a CpG dinucleotide (lower map) or a GpC dinucleotide (upper map). Generally the distribution of the dinucleotide CpG is depleted in the genome. However, CpG-rich islands are often found at the 5′ end of genes. Note that CpG-rich implies a region where the frequency is similar to that of GpC dinucleotides, which are not depleted in the genome.

constructed from the mRNA from muscle might be screened. Any positive cDNA, representing an mRNA expressed in muscle, which maps to the same area of the genome as the original linked markers would be a candidate for the gene in question.

12: Analysis of cloned DNA molecules

Sequencing DNA molecules

A number of techniques for determining the nucleic acid sequence of a DNA clone have been designed. Currently, the most frequently used is the method of Sanger and his coworkers (Fig. 12.1). This method requires a single-stranded DNA template, so the DNA molecule to be sequenced is first cloned into the vector bacteriophage M13. As has been described above, this vector makes possible the synthesis of a single-stranded phage DNA molecule containing the cloned DNA as an insert. The DNA in M13 phage particles is single-stranded but during the intracellular part of its life cycle M13 DNA becomes double-stranded. It is possible to isolate double-stranded M13 from within the host bacteria. This is now available commercially. The double-stranded foreign DNA is ligated into the cloning site of the double-stranded M13 in the same way as in plasmid cloning. The M13 and the foreign DNAs are first digested with a restriction enzyme and are then ligated together. Some of the M13 molecules will circularize in the ligation reaction, and some of these will contain a foreign DNA insert. The M13 is introduced into the host bacteria by heat-shock transformation as with plasmids. The transfected cells, plated out on to agar, will then synthesize M13 phages, visible as plaques. The plaques containing recombinants can be distinguished from non-recombinants by colour selection. Plaques containing recombinants are picked and phage grown from them. The single-stranded DNA from within the phage particles is isolated. An oligonucleotide primer is annealed to its complementary sequence on single-stranded M13, 20–40 bases downstream (3′) of the cloning site. Four reactions are then performed in which a DNA polymerase synthesizes a complementary strand beginning at the priming site and running through into the clone. Each reaction contains the template, the enzyme and all four single nucleotides needed for second-strand synthesis, one of which is radiolabelled. In each of the reactions there is also, in addition to the normal deoxy version (dNTP), a dideoxy version (ddNTP) of one of the four nucleotides. The four reactions therefore contain ddT, ddA, ddG and ddC respectively, as well as dA, dT, dC and dG. Consider the reaction containing ddG. Each time the polymerase reaches a C on the template, it will incorporate a G into the second strand. Usually this will be a dG and synthesis of the second

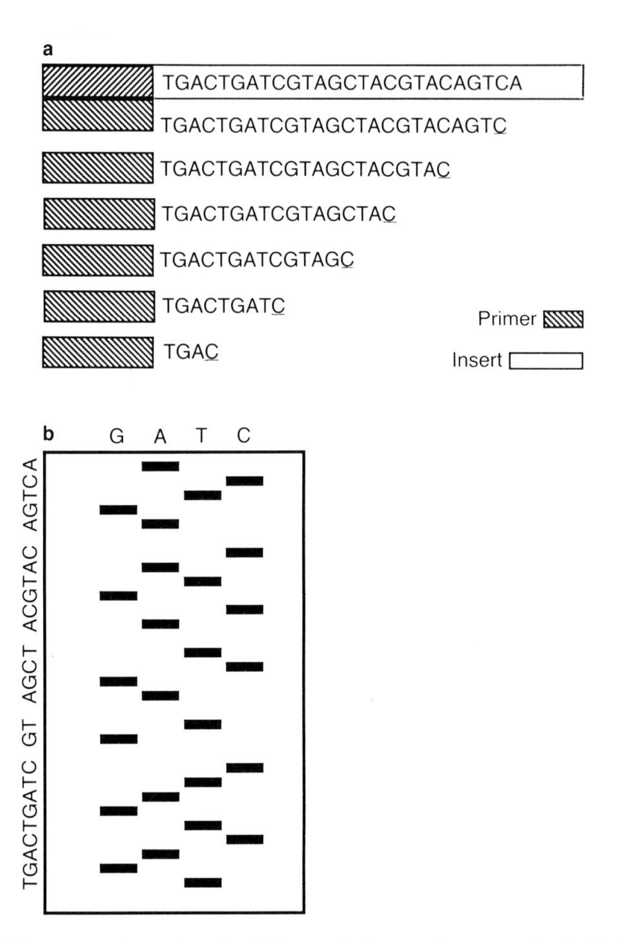

Fig. 12.1 DNA sequencing using the dideoxy chain terminator method. (a) A primer is annealed to the single-stranded M13 template containing the insert. Four reactions are performed each containing all four single nucleotides, one of which is radiolabelled, and one dideoxynucleotide. DNA polymerase synthesizes a complementary strand until a dideoxynucleotide is incorporated ending extension. The ratio of deoxynucleotides to dideoxynucleotides is adjusted so that a full range of lengths is produced. (b) The newly synthesized DNA from each of the four reactions is separated by gel electrophoresis and the sequence read.

strand will continue. In some cases, however, a ddG will be incorporated. Since ddG cannot form phosphodiester bonds with the next incoming nucleic acid, no further bases can be added on to it and the second strands will stop being synthesized. The ratio of dG to ddG is carefully controlled so that ddG is incorporated sufficiently frequently for a range of second-strand molecules of all sizes to be produced, each terminating in a G base. Since shorter molecules each carry fewer radiolabelled bases than longer molecules and the aim is to produce an autoradiograph on which each molecular size is equally visible, under ideal circumstances rather more smaller molecules will be synthesized than longer molecules. The same events take place in the ddT, ddA and ddC reactions. The resulting second-strand fragments are then separated from the template by heat denaturation and fractionated on a denaturing polyacrylamide gel. After autoradiography a characteristic ladder pattern is produced from which the sequence can be read (Fig. 12.2).

Sequence analysis using computers

Once the sequence of a DNA molecule of interest has been identified, it may be analysed using one of a number of specialized sequence analysis

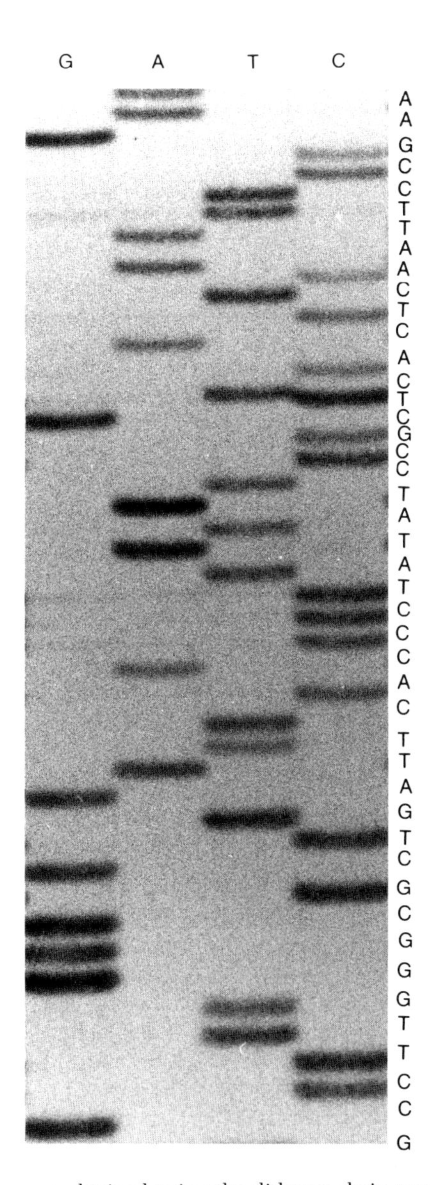

Fig. 12.2 DNA sequence obtained using the dideoxy chain terminator method.

software packages. Those which run on personal computers are limited to identification of open reading-frames, translation to protein sequence, restriction site identification and other simple manipulations of the sequence itself. Programs which run on mainframe computers allow more complex and time-consuming analysis such as nucleic acid and protein homology searching, searching for possible coding regions and constructing 'contigs' of several related sequences. Although mainframe computers running these programs are available in some institutions, most groups will access a suitable mainframe machine by telephone link. In the UK, the Joint Academic Network for Electronic Transmissions (JANET) allows communication with the DNA analysis computer at the Science and Engineering Research Council (SERC), Daresbury, which runs all of the major sequence analysis packages and has access to DNA databases from several sources.

SEQUENCE ANALYSIS ON A PERSONAL COMPUTER

Sequence analysis programs for personal computers generally accept input of the sequence from the keyboard or from files derived from programs for reading sequencing gels (Fig. 12.3). In some cases a proof-reading facility is available which reads the sequence back to the operator using the computer's sound facilities. The sequence can then be translated in each of the six possible frames. There will be three possible frames on the single-strand sequence entered and three more on the complementary strand. Since the orientation of the genomic fragment that has been sequenced is probably not known, any genes within it may be on either of the complementary strands. The program will search each of the six frames for an open reading-frame and

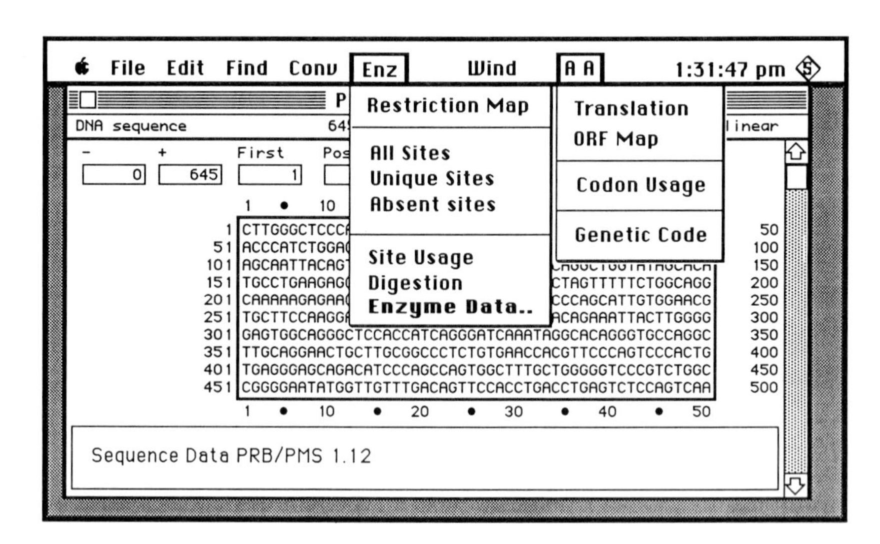

Fig. 12.3 Screen shot from the Strider DNA analysis package designed to run on an Apple Macintosh personal computer.

give the protein translation of the sequence. Some programs will then allow further analysis of the protein sequence, giving information about secondary structure, hydrophobicity, transmembrane portions and antigenic index.

A personal computer-based sequence analysis package will usually also allow restriction enzyme site analysis. The program will locate restriction enzyme recognition sites within the sequence. Output can be requested which shows all recognition sites, only recognition sites for commonly used enzymes or only recognition sites which are unique or occur a specific number of times within the sequence. From this it would be possible to predict restriction fragment sizes and to select suitable strategies for subcloning.

SEQUENCE ANALYSIS ON A MAINFRAME COMPUTER

A mainframe computer has the advantages of greater memory, a faster processing speed and access to large, regularly updated DNA databases. Many of the operations which are usually performed on a mainframe computer may now be performed with the new generation of powerful software for personal computers. Personal computers may have access to DNA databases on optical disc but these are expensive and have the significant disadvantage that they are not updated as regularly as the large institutional mainframe machines. In addition, a mainframe machine often has access to several different DNA databases.

The DNA databases available include NBRF (National Biological Research Foundation, USA), Genbank (Genetic Sequence Data Bank, BBN Laboratories, USA) and EMBL (European Molecular Biology Laboratory, Germany). These databases are similar to each other although there are variations in the style of their files. DNA sequences are submitted by investigators from all parts of the world as they become available. In many cases an individual sequence from one investigator will appear on all of the databases but in other cases it will appear on only one of them. Databases are also available containing sequence information for cloning vectors (Vecbase, Max Planck Institute, Germany) and there are analogous databases containing protein sequence data.

Mainframe computers running DNA analysis software such as Staden or the University of Wisconsin packages will perform all of the analyses available on personal computer-based packages. In addition, they allow more sophisticated analysis. For example, a series of files containing overlapping sequence data, possibly a series of cosmid clones, can be automatically assembled into a single contiguous sequence (a 'contig') even when there are discrepancies in the sequence data or portion of vector sequence present. The program will then also determine the most likely correct sequence and produce an overall 'consensus' sequence.

Large consensus sequences can be analysed to identify probable coding sequences based on nucleic acid frequency. The method of 'uneven positional base preference' is based on the assumption that, in a DNA sequence which does not code for protein, the proportion of each base in any given codon position will be the same as for the proportion of that base in the sequence overall. No preference for any particular type of base for any particular position will be seen. In a DNA sequence which does code for a protein, there will be a particular preference for some bases in particular codon positions. For example, one base may appear more frequently in the first codon position than would be expected from its overall proportion in the sequence. A value indicating how far the bases of each type are from being evenly distributed amongst the codon positions of an arbitrary reading-frame is calculated. The greater the value, the greater the probability that the sequence does code for a protein.

The DNA databases can simply be searched, using keywords, to discover whether a particular gene has been cloned and to find its sequence. They may also be used to determine homology between a new sequence and sequences in the database. Homology searching is performed using a technique known as dot matrix plotting. The program lays the sequences to be compared along vertical and horizontal axes. A window of fixed, odd, length is laid at the start of each sequence. The sequences within the windows are compared and, if they are sufficiently similar, a dot is placed at the position of intersection of horizontal and vertical lines drawn from the middle of each window. How similar the two sequences need to be for a dot to be drawn can be determined by the operator. The horizontal window is then moved one base to the right and the new sequences compared. This is repeated until the horizontal window has moved the full length of the sequence on the horizontal axis. The vertical window is then moved up one base and the whole procedure is repeated. Eventually every possible window-sized subsequence of the vertical sequence will have been compared with every possible window-sized subsequence of the horizontal sequence. A series of dots will have been plotted showing where the sequences are similar (Fig. 12.4). Where the overall sequences have homologous regions, this will be seen as a series of dots. The longer and more contiguous the lines of dots, the greater the homology between the two sequences. The program will perform homology analysis simply between two sequences or will compare a single sequence with every sequence in one of the DNA databases. This procedure, even for a mainframe computer, may take several hours.

Identification of mutations causing the disease phenotype

Once a candidate gene has been isolated and its sequence analysed, it is necessary to determine how the gene of interest differs in the disease state from the normal state. If the mutation is a deletion of sufficient size, this can be identified initially by Southern blotting. However, if the mutation is only a minor alteration to the DNA sequence (e.g. a

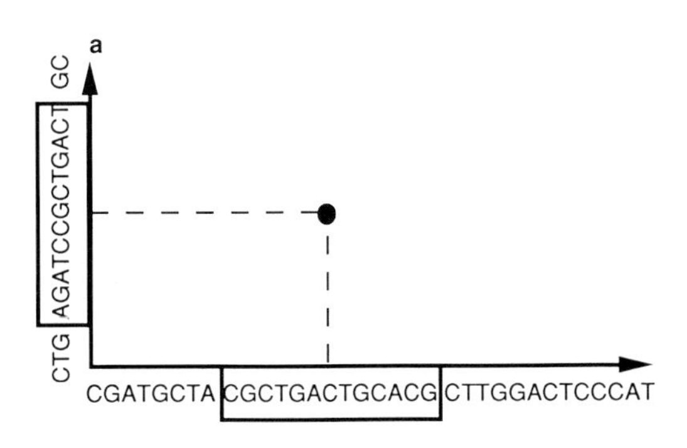

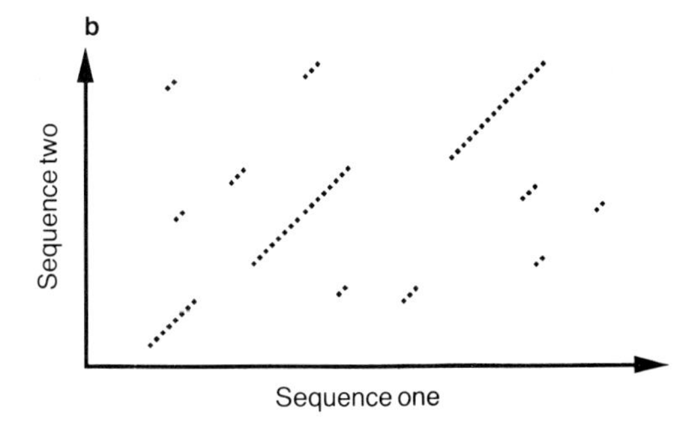

Fig. 12.4 DNA sequence homology searching, using dot matrix plotting on a mainframe computer. (a) Sequences are compared along horizontal and vertical axes and are point-plotted where homology is found. (b) When the analysis is completed, regions of homology are seen as diagonal lines of dots.

single base pair substitution), there will be no clues as to where to look and the whole gene sequence will need to be compared between normal and affected individuals. Several years ago this proved to be laborious, since to obtain clones from different individuals required making cDNA or genomic DNA libraries, isolating the clone(s) of interest and sequencing them. Recently, however, a number of new strategies have been developed to make this process easier. These strategies rely on PCR amplification to enable individual specific gene sequences to be readily available. The second stage, comparison of the DNA sequence between normal and affected individuals, can be carried out in several different ways, the most important of which are systematic sequencing of the PCR products (direct sequencing), denaturing gradient gel electrophoresis (DGGE) and chemical cleavage analysis of heteroduplexes using hydroxylamine and osmium tetroxide (HOT technique).

DIRECT SEQUENCING

Direct sequencing simply requires sequences to be obtained from one end of the gene to the other in normal and affected individuals. The sequences are compared to identify any differences. Sequencing can be performed direct from PCR products without the need to subclone the fragments into bacteriophage M13. Once the gene fragment has been amplified Sanger sequencing can be performed using a primer internal to the original primers used in the PCR. Only one primer, complementary to one strand, is used. Problems arise with this technique when the gene sequence is large, as it is not easy to sequence more than about 200 bases from one sequencing primer. But, as with M13 sequencing, the length of sequence data obtained can be extended by synthesizing new primers further along the strand when that part of the sequence is known. The quality of sequence obtained by direct sequencing is poor compared with sequencing using M13 and often gives rise to ambiguities.

DENATURING GRADIENT GEL ELECTROPHORESIS

The method of DGGE is based on the differential melting of DNA when electrophoresed through a gel containing an increasing concentration of denaturant. As the DNA duplex migrates into a portion of the gel containing sufficient denaturant concentration, the DNA strands separate, retarding further progress through the gel. Different DNA sequences of the same length will denature at different distances on the gel, depending on their specific base composition and the order of the bases. It is possible to differentiate two DNA duplexes containing no mismatches and differing only by a single base using this method. However, greater sensitivity can be achieved when one of the duplexes contains a single-base mismatch — a heteroduplex. This can be achieved either by annealing PCR products obtained from a normal individual with an affected one or by analysing a heterozygote for the disease. To form heteroduplexes, it is necessary to denature the PCR products and allow them to reanneal. This method is a sensitive way of identifying sequence differences but does not give the exact location or precise nature of the difference. This then has to be further characterized by sequencing.

HYDROXYLAMINE AND OSMIUM TETROXIDE TECHNIQUE

The hydroxylamine and osmium tetroxide (HOT) technique can be used to locate precisely mutations, although confirmatory sequencing is also required to characterize the exact nature of the mutation. This method relies on the formation of heteroduplexes either from heterozygotes or by mixing affected and unaffected DNA PCR products. The PCR products are then split into two to react with hydroxylamine and osmium tetroxide. Hydroxylamine reacts with mismatched cytosine residues and osmium tetroxide reacts with mismatched thymine residues. When these products are then reacted with piperidine, the DNA strand is cleaved. Analysis on a polyacrylamide gel reveals the cleavage products from these reactions. Cleavage with these two chemicals allows all mismatches to be detected. By over-reacting the PCR products in a time course, a cytosine or a thymine sequencing ladder can be produced, indicating the position of the mutation.

Once a mutation has been found, it is important to check that it will cause some change to the protein sequence such as a frame shift or an amino acid substitution. Minor changes may simply be silent polymorphisms. It is important to follow up findings with further evidence before being satisfied that the disease mutation has been identified. For example, one would expect to find the mutation to be homozygous in all individuals with the disease, heterozygous in carriers of the disease and not present in normal individuals.

13: Studying the expression of genes

A gene is expressed when it is transcribed into mRNA and that mRNA is translated into protein. Although gene expression can be studied by studying protein production, there are other biochemical factors which may affect protein synthesis and, when a suitable probe is available, it is often more appropriate to study the synthesis of the specific mRNA.

Isolation of RNA and northern blot analysis

One of the simplest ways of studying gene expression is by northern blot analysis. A northern blot is very similar to a Southern blot except it is RNA rather than DNA which is blotted. The RNA is extracted from

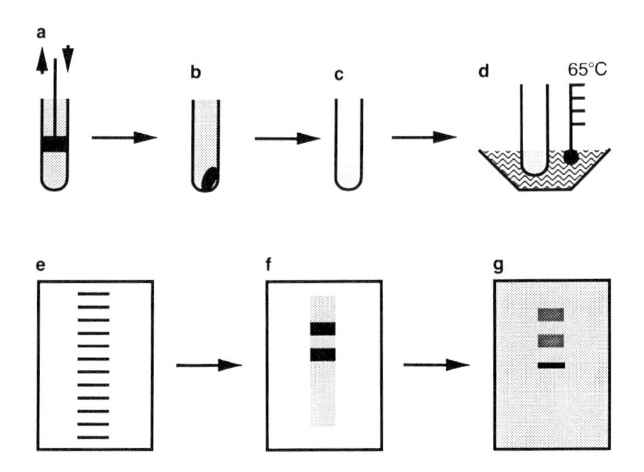

Fig. 13.1 RNA isolation and northern blotting. (a) The tissue or cells are homogenized in a denaturing solution containing guanidinium and phenol. (b) The RNA is separated by ethanol precipitation followed by centrifugation. (c) The RNA is redissolved in RNase free water. (d) The RNA is heated to destroy secondary structure and (e) the molecules are separated by electrophoresis in a gel containing a denaturing agent such as formaldehyde. Small RNA molecules travel further down the gel than larger molecules. (f) After electrophoresis RNA appears as a streak representing molecules of a wide range of sizes. Usually two denser areas are seen representing the ribosomal RNA. The RNA can be transferred to a nylon filter in the same way as in Southern blotting. This is termed northern blotting. (g) Hybridization with a specific probe detects a specific mRNA type. There may also be non-specific hybridization to the ribosomal RNA.

the tissue or cells being studied and separated by electrophoresis on an agarose gel (Fig. 13.1). Since RNA is readily degraded by RNases found almost everywhere, great care needs to be taken. All glassware is baked and plasticware chemically treated to destroy RNases. Gloves are worn since the skin contains RNases. The tissue or cells are either snap-frozen or placed into a denaturing solution as soon as possible to prevent degradation. There are various methods for extracting RNA from tissue but all are based on similar principles. The tissue is homogenized in a denaturing solution containing guanidinium, phenol and chloroform. The guanidinium is a strong denaturant which destroys proteins present in the tissue, including RNases. The phenol and chloroform will dissolve proteins and lipids and after centrifugation the nucleic acids remain in the aqueous phase whilst the protein and lipids are in the phenol phase. The RNA can then be separated by careful adjustment of the precipitation conditions, using ethanol. After a second centrifugation, the RNA, which appears as a white solid, can be seen at the bottom of the tube. The RNA is then resuspended in RNase-free water.

RNA is single-stranded but often forms a secondary structure by folding and hybridizing to complementary regions within the RNA molecule itself. This is prevented by heating to 65°C for 10 minutes before electrophoresis and by using gels containing a denaturant such as formaldehyde. Once the gel has run, the RNA can be visualized under ultraviolet light after staining with ethidium bromide. RNA appears as a streak, much like digested DNA, but also visible at a migration

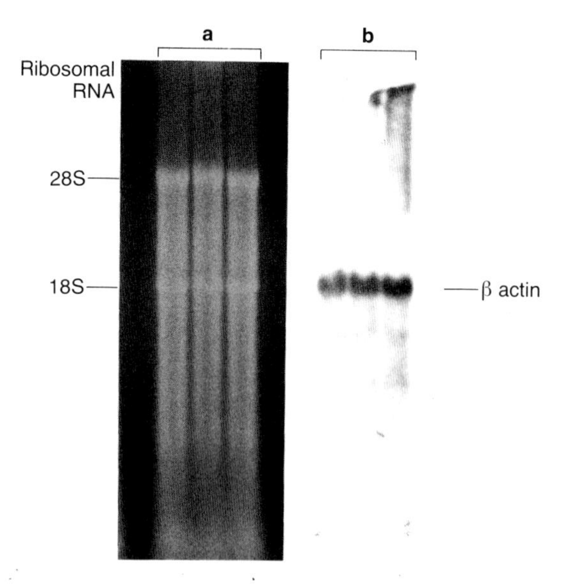

Fig. 13.2 (a) RNA separated on a 1.5% denaturing agarose gel. The visible bands represent ribosomal RNA whilst messenger RNA appears as a streak. (b) The gel has been northern-blotted and the filter hybridized to a cDNA probe for β-actin.

distance equivalent to about 4 and 2 kb are denser bands. These represent the 28S and 18S ribosomal RNA (Fig. 13.2a). Ribosomal RNA represents over 90% of total RNA. The degree to which the RNA has been degraded can be assessed from the appearance of the ethidium-stained gel. Undegraded RNA appears as an even streak running the length of the gel with clearly defined ribosomal bands. Occasionally other faint bands can be seen which represent single mRNA species in high copy number, and the 5S ribosomal RNA may be seen as a band nearer to the positive electrode. Degraded RNA is of smaller molecular weight, so that the streak appears to be shifted towards the end of the gel. In gels containing degraded RNA, ribosomal bands are not as distinct and may appear to be of smaller molecular weight or may even be absent.

RNA can be transferred to a nylon membrane by a method similar to Southern blotting — hence the term northern blotting — and can be hybridized to a probe to detect a specific mRNA species. Double-stranded DNA probes, e.g. cDNAs, can be used, although only the antisense strand will hybridize to the single-stranded RNA. RNA probes (riboprobes) can also be used. These are synthesized using a vector containing the promoters for one or more RNA polymerases. Again only the antisense RNA probe will work. Northern blot analysis gives information about the presence or absence of a specific mRNA and about its molecular size. It does not provide information about the precise concentration of an mRNA species, although some idea about abundance can be gained by comparing the intensity of hybridization with the hybridization of a probe specific for a known abundant protein such as β-actin. In cases where northern blot analysis suggests that there is no expression of a particular gene, rehybridization with β-actin will confirm that the RNA itself was in good condition and that the hybridization reagents are working (Fig. 13.2b). Usually it is total RNA which is loaded on to a gel for northern blot analysis; however, it is sometimes appropriate to load poly-A-enriched RNA obtained using oligo-dT chromatography (see Chapter 10). This is especially the case if the probe being used hybridizes non-specifically to ribosomal RNA, if the transcript size is similar to that of ribosomal RNA (4.2 and 1.8 kb) or if the mRNA is in very low copy number.

It is possible to detect the presence of an mRNA species using dot blots or slot blots, as can be done with DNA. Although these techniques are much quicker than Southern blotting, they suffer from a number of disadvantages when compared with northern blot analysis. In a total RNA dot blot, over 90% of the RNA will be ribosomal. Many probes will hybridize non-specifically to ribosomal RNA and, unless the mRNA being detected is in very high abundance, non-specific hybridization may mask any differences in expression. Unlike DNA, RNA is very prone to degradation. A dot blot does not allow any assessment of the integrity of the RNA, and apparent absence of expression may be due to unseen degradation. Despite these disadvantages, some experimenters

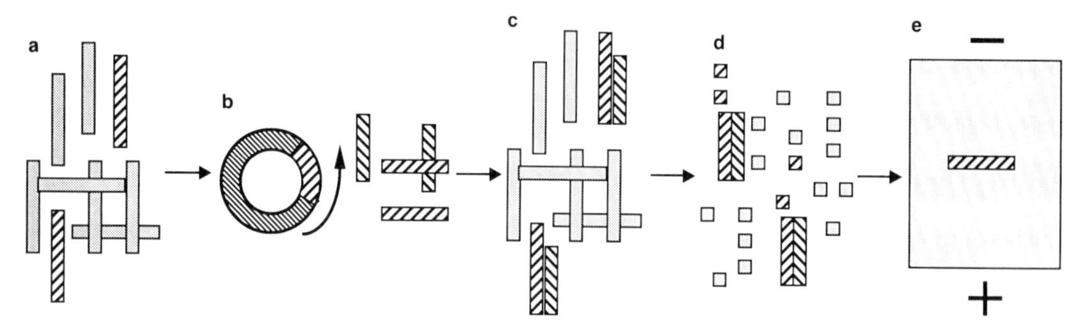

Fig. 13.3 Ribonuclease protection assay. (a) RNA is isolated from the tissue or cells being studied. (b) A cDNA for the mRNA species to be detected is cloned into an expression plasmid which allows synthesis of single-stranded RNA. Antisense RNA is synthesized. Inclusion of a radiolabelled nucleotide in the reaction produces radiolabelled (RNA) riboprobes. (c) The riboprobes are mixed with the total RNA and hybridize to the specific mRNA species to be detected. (d) Treatment with RNase destroys both the original RNA and any unhybridized riboprobe. The specific mRNA hybrid is protected and remains intact. If the riboprobe is shorter than the mRNA then the overhang will also be destroyed. (e) The mRNA riboprobe hybrid can be separated from the single nucleotides by electrophoresis and identified by autoradiography. Alternatively it may be isolated by column chromatography and its concentration calculated by counting its radioactivity.

continue to use RNA dot or slot blot techniques. Great care needs to be taken in the interpretation of their results.

Ribonuclease protection assay

Ribonuclease protection assay gives better information about the actual concentration of an mRNA species (Fig. 13.3). In this case an antisense riboprobe is synthesized and hybridized *in vitro* to the total RNA population. The riboprobe will anneal only to the specific mRNA to be detected. The RNA is then treated with an RNase which destroys only single-stranded RNA, leaving the mRNA riboprobe hybrid intact. Any unhybridized riboprobe will also be destroyed. Since the riboprobe is originally radiolabelled, the remaining double-stranded hybrid can be detected by autoradiography and its concentration calculated from the degree of radioactivity present.

In situ hybridization to RNA

In situ hybridization allows identification of the sites of gene expression on a histological section. The tissue to be examined is fixed or frozen-sectioned, using techniques which protect against RNA degradation. It can then be hybridized to either radiolabelled riboprobes or DNA oligonucleotide probes to detect the presence of the mRNA species of interest. After autoradiography, silver grains on the specimen indicate

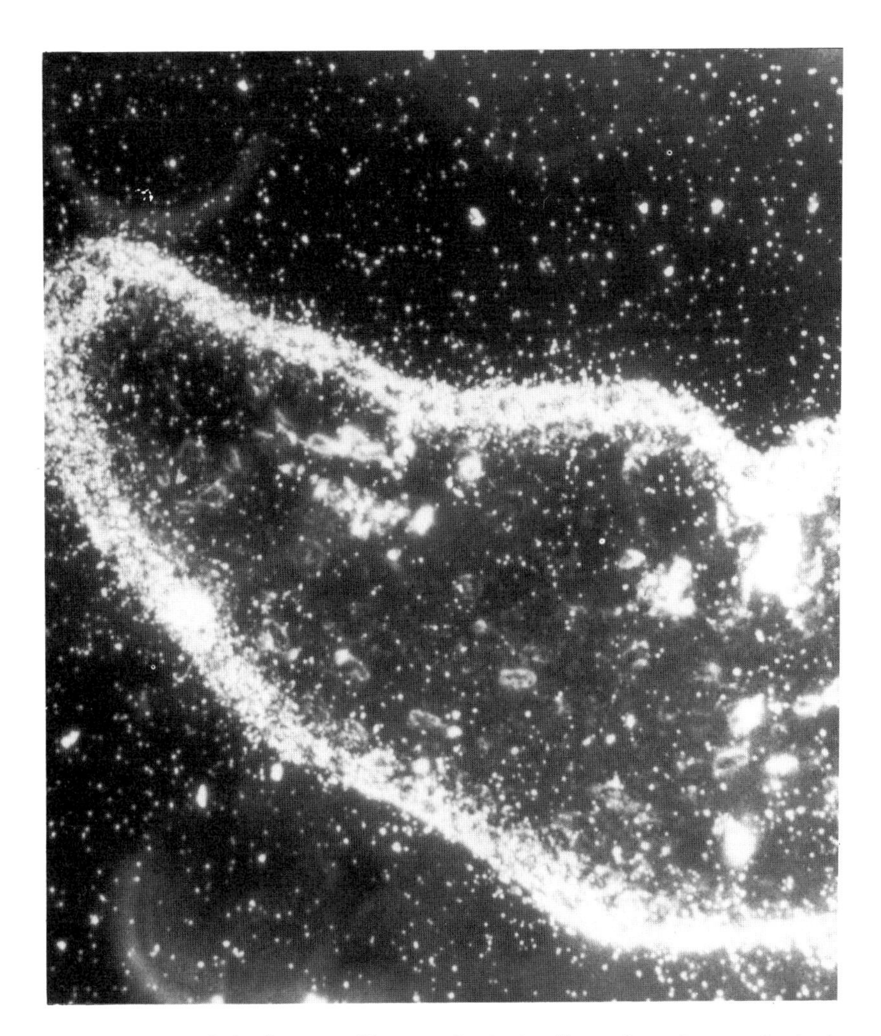

Fig. 13.4 *In situ* hybridization of human chorionic villus using a human chorionic gonadotrophin probe labelled with ^{35}S. Microscopy is dark-field with objective magnification of ×40. Hybridization is restricted to the outer syncytial trophoblast layer with few silver grains visible over the inner cytotrophoblast layer.

the presence of the hybridized probe (Fig. 13.4). Only antisense probes will hybridize to the mRNA; sense probes may be used as a control to assess the degree of non-specific hybridization.

Part 3
The Clinical Applications of Molecular Biology

14: Prenatal diagnosis

A major application of molecular biology to obstetrics and gynaecology is in the prenatal diagnosis of inherited disease. The approach taken will depend upon what is known about the molecular pathology of the disease in question. In many cases the DNA mutation is known and can be detected in DNA obtained from the fetus. In other cases, whilst the mutation itself has yet to be found, linked DNA polymorphisms can be used to determine whether or not the fetus has inherited the disease. In some cases linkage analysis is preferable to direct gene detection, particularly when a single disease may be caused by one of several mutations, as is the case, for example, in β-thalassaemia. In the case of X-linked diseases for which there is neither linkage nor information about the mutation, DNA technology can be used to sex the fetus and allow termination of pregnancy in the case of a male. This is clearly an unsatisfactory situation, which will become less common as more information becomes available about these diseases. Perhaps more acceptable is the preimplantation sexing of embryos conceived *in vitro* and replacement of only females, although again, as DNA technology improves, it may be possible to identify single-gene defects and select unaffected males.

At the biochemical level there are essentially three approaches to the identification of a molecular pathology. In cases where the mutation creates or destroys a restriction site or where there is a large deletion or insertion the diagnosis may be made by restriction mapping. If the defect causes no change in restriction fragment sizes the individual mutation may need to be examined directly. Lastly, and often the most convenient approach is linkage analysis which can be used to determine whether the fetus has inherited the mutation without the need to analyse the gene itself.

Restriction mapping

Major deletions or insertions will change the distance between restriction sites and can be detected by restriction mapping. Alpha-thalassaemia, for example, is frequently caused by a large deletion in the α-haemoglobin gene. All of the haemoglobins have a tetrameric structure. In adult and fetal haemoglobins two of the chains are always

α. The type of haemoglobin is determined by the type of chain linked to the α chains. Adult HbA has β chains and adult HbA2 has δ chains. Fetal HbF has γ chains. There are two types of γ chain, differing only by a single amino acid, glycine or alanine at position 136. HbF is a mixture of the two types. Embryonic haemoglobin may have either α or ζ chains combined with either γ or ε.

The α and ζ genes are close together on chromosome 16. There are two α genes: α_1 and α_2. Just upstream from these are two pseudogenes, $\psi\alpha$ and $\psi\zeta$. These have homology to their functioning counterparts but were presumably disabled by mutations at some time during evolution. The ζ gene is just a little further upstream. Similarly the β genes are close together on chromosome 11 in the order 5′ ε $^{G}\gamma$ $^{A}\gamma$ ψ β δ β 3′. The gene $\psi\beta$ is also a pseudogene. The α gene family all have an identical intron arrangement, as do the β family, since each family was formed by a series of duplication events (see Chapter 3).

Alpha-thalassaemia is caused by reduced synthesis of the α chain of haemoglobin. There are two α genes on each chromosome. In some cases only one of the genes is mutated and the other functions normally. This is α^+ disease. In other cases both of the genes on one chromosome are mutated, producing the α^0 disease. Homozygous inheritance of α^0 means that no α-haemoglobin is produced. Prenatally the γ chains combine together to produce HbBarts ($\gamma4$) and with the ζ chains to produce HbPortland ($\zeta2\gamma2$). There is intrauterine hypoxia and the infant is usually stillborn or dies in the early neonatal period. Homozygous inheritance of α^+ is compatible with survival since some α chains are produced.

The α^0 thalassaemias are caused by large deletions which may span both of the α genes. The deletion usually begins in the α_1 gene and may include part or all of the α_2 gene and sometimes the adjacent pseudogenes. If the deletion does not remove both of the α genes, it can be detected by using the restriction enzyme *BamHI* and a gene probe for the α-globin gene. The normal fragment size is 14 kb but with one of the common deletions (4 or 3.7 kb) a smaller fragment of 11.3 or 10 kb will be seen. Larger deletions will remove both of the α genes. In this case the α probe will not detect any band. If the individual is heterozygous, the band from the normal chromosome will be of normal size and could be mistaken for a homozygous normal. In this case the diagnosis may be made by using a ζ gene probe together with the restriction enzyme *BglII*. The normal band size in this case is variable, since there is normally a region of hypervariability near to the ζ gene, but it is usually within the range 9.5 to 12.5 kb. Deletions may produce smaller bands or, because they delete a recognition site for *BglII*, bands which are larger than normal. Rarely, a deletion is so large that it removes all of the genes in the α gene cluster, and diagnosis then requires a probe which detects a region of DNA near to the gene cluster.

There are many different molecular pathologies which cause β-

thalassaemia, most of which are point mutations or small deletions that cannot be detected by restriction mapping. There is one form of β-thalassaemia in certain Indian populations which is due to deletion of 619 bases at the 3′ end of the β-globin gene. This deletion shortens the distance between two *BglII* sites from 5.2 to 4.6 kb. After digestion with *BglII*, Southern blotting and hybridization to a β-globin probe DNA from a homozygote-affected individual will display only a 4.6 kb band. A heterozygote will display both 5.2 and 4.6 kb bands whilst normals have only a 5.2 kb band (Fig. 14.1).

The mutation which causes sickle-cell disease is a point mutation which can be detected by restriction mapping. Sickle-cell haemoglobin is produced by a single-base substitution, A for T in the sixth codon of the β-globin gene. This changes the codon from GAG, coding for glutamic acid, to GTG, coding for valine. The surrounding normal sequence CCTGAGGAG is recognized by the restriction enzyme *MstII*. The mutation to CCTGTGGAG destroys this recognition site, therefore, if the DNA from a homozygous sickle-cell sufferer is digested with *MstII*, Southern-blotted and hybridized to a β-globin probe, a fragment will be produced which is 0.2 kb larger than the normal 1.1 kb. A heterozygote would produce one 1.1 kb and one 1.3 kb fragment (Fig. 14.2).

There are a number of other genetic diseases, due, for example, to deletions in the antithrombin III and blood clotting factor VIII and IX genes, which can be detected in this way. In many other cases, the mutation does not affect the restriction map and so mutations need to be detected directly.

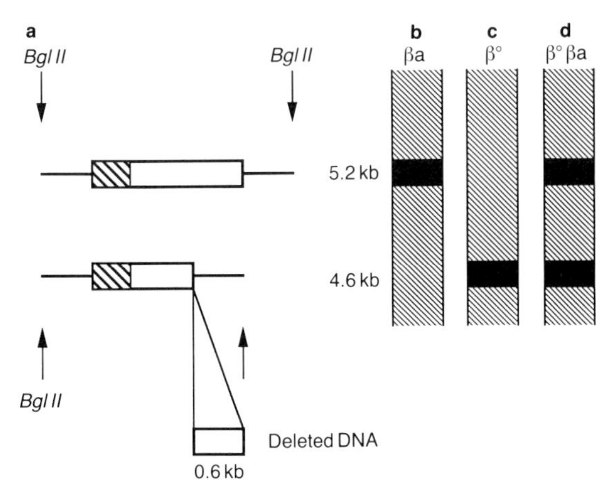

Fig. 14.1 Restriction analysis of β-thalassaemia. (a) Deletion of a 0.6 kb portion of the gene shortens the distance between two *BglII* sites. (b) A β-globin-specific probe detects a 5.2 kb fragment in normal individuals and (c) a 4.6 kb fragment in individuals homozygous for the deletion. (d) In heterozygous individuals both fragments will be detected.

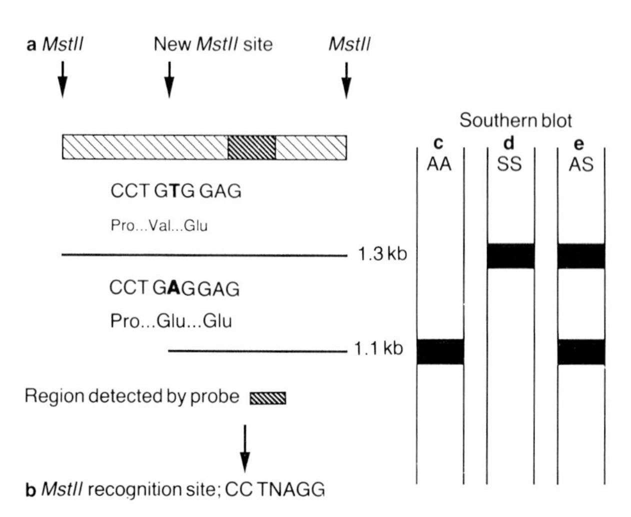

Fig. 14.2 Detection of sickle-cell disease using restriction analysis. (a) The point substitution **A** for **T** changes the amino acid coded from glutamic acid to valine and creates (b) a recognition site for the restriction enzyme *MstII*. A β-globin-specific probe detects (c) a 1.1 kb fragment in normal individuals and (d) a 1.3 kb fragment in individuals homozygous for the deletion. (e) In heterozygous individuals both fragments will be detected.

Direct detection of point mutations

In general, probes used in Southern blotting are large DNA probes, derived from cDNA or genomic libraries, and may be several hundred to several thousand base pairs in size. These probes will detect not only regions to which they have perfect homology but also regions where homology is less than 100%. For example, a gene probe for the mouse cyclo-oxygenase gene will detect the human gene, although there is less than 80% direct homology between them. The hybridization tends to remain stable over a wide range of stringency, although, if the temperature of hybridization or post-hybridization washing is increased enough and the salt concentration decreased sufficiently, hybridization to regions of lower homology is reduced. It would, however, be completely impossible for a DNA probe of, say, 1 kb to detect the difference in homology due to a single-base change. This is not the case for a DNA probe consisting of only a few base pairs — an oligonucleotide probe — for which a single-base change may radically alter the hybridization conditions.

Hybridization of oligonucleotide probes is very dependent upon temperature and hybridization conditions. An approximation to the maximum hybridization temperature can be calculated from the base composition of the probe using the formula:

$$2(A + T) + 4(G + C)$$

but the exact conditions needed to detect a single-base change will need to be determined by trial and error.

For diagnosis of a single-base change, two oligonucleotides of about 20 base pairs are synthesized, one identical to the normal sequence (N) and one identical to the mutated sequence (M). M will therefore be the same as N except for the single-base change. Both the test DNA and a sample of known normal DNA are digested with a restriction enzyme and two Southern blots are made, each containing both DNA samples. One of these is hybridized to the 'normal' oligonucleotide (N) whilst the other is hybridized to the mutated oligonucleotide (M). The N oligonucleotide will hybridize to the normal DNA but not to the abnormal, whereas the M oligonucleotide will hybridize to the abnormal and not to the normal DNA (Fig. 14.3).

Using polymerase chain reaction in prenatal diagnosis

As discussed in Chapter 9, the polymerase chain reaction (PCR) is a method for amplification of a specific region of DNA from as little as one copy to produce sufficient DNA for it to be visualized on an agarose gel stained with ethidium bromide. In general, the amplification

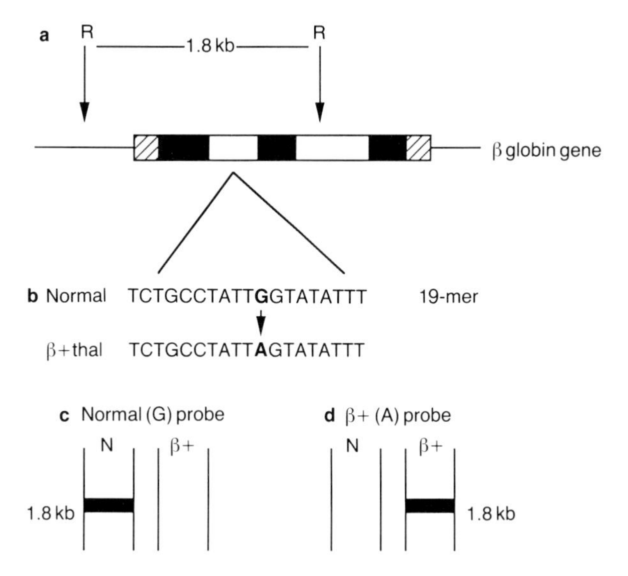

Fig. 14.3 Use of an oligonucleotide probe to detect a point mutation which causes β-thalassaemia. In accordance with the β-globin gene (a), two oligonucleotides are synthesized (b), one of the correct sequence (TCTGCCTATTGGTATATTT) and the other of the sequence with the point mutation (TCTGCCTATT**A**GTATATTT). Genomic DNA is digested with a suitable restriction enzyme, the fragments are separated by electrophoresis and duplicate Southern blots are made. One Southern blot is hybridized to the 'normal' oligonucleotide (c) and the other to the β-thalassaemia oligonucleotide (d), under conditions which only allow annealing when there is perfect homology. In the normal homozygote only the 'normal' oligonucleotide will anneal, in the β-thalassaemia homozygote only the β-thalassaemia oligonucleotide will anneal. In the case of a heterozygote both oligonucleotides will anneal.

reaction can be done unattended and automatically, and the 'results' can be seen on an agarose gel. In many applications, PCR eliminates many of the time-consuming aspects of DNA diagnosis, e.g. Southern blotting, and some of the more hazardous aspects, such as the use of radiolabelled probes. The greatest advantage that PCR has over more traditional techniques is that it requires such small amounts of starting template. So, for example, it could be used to identify carrier status for a single-gene disorder using cells from a buccal scrape and mouthwash. It has also been used to amplify DNA from the leucocytes in a Guthrie spot taken from a neonate who subsequently died. The ability to amplify DNA from a single cell has led to experiments in DNA diagnosis in embryos fertilized *in vitro* prior to implantation.

Large deletions are not conveniently detected by PCR. A suitable strategy would be to make oligonucleotide primers to regions either side of the deletion and to amplify across it. Unfortunately PCR is less successful when amplifying regions of more than a few kilobases. In the case of α-thalassaemia, for example, the 'normal' fragment might be up to 15 kb in size. Until recently, it was also difficult to use PCR to detect small deletions. A difference of 1 or 2 base pairs in a fragment of 100 base pairs is difficult to detect using agarose gel electrophoresis. However, refinements in electrophoresis techniques have made separation of fragments differing by only a few base pairs possible. A good example is the diagnosis of cystic fibrosis. In 70% of cases, the cystic fibrosis mutation is a deletion of a phenylalanine codon at position 508

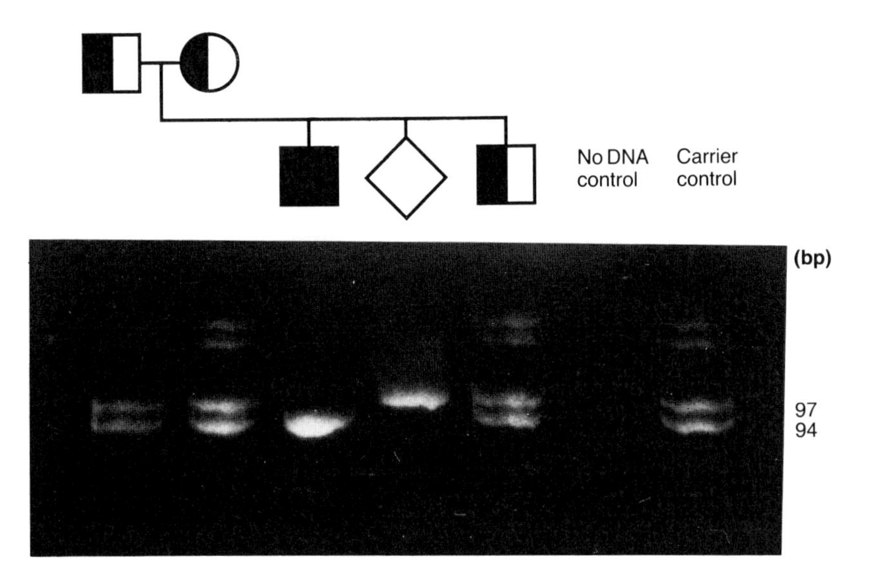

Fig. 14.4 PCR diagnosis of cystic fibrosis by direct analysis of the △508 deletion. On a 12% polyacrylamide gel PCR products that are 3 bp different in length can be separated. The faint products visible in the parents' and the control lanes result from heteroduplexes formed between the 94 and the 97 bp products. The mismatched region causes the resulting heteroduplexes to run more slowly through the gel.

of the gene. This phenylalanine deletion is in a potential ATP-binding site in the protein. PCR primers have been made which amplify a 100 base pair region flanking amino acid position 508. In the presence of the deletion, the product formed is 3 base pairs shorter. These two fragment sizes can be resolved using a 12% polyacrylamide gel and can be seen by staining the gel after electrophoresis with ethidium bromide (Fig. 14.4). Of course, this method will not necessarily detect one of the other cystic fibrosis mutations, e.g. a single-base change. Although most cases of cystic fibrosis in Britain are due to phenylalanine deletion at position 508, it is now apparent that there may be a large number of rarer mutations which also cause cystic fibrosis.

Another method for identifying very small deletions or single-base changes is the technique of heteroduplex formation. PCR is performed using the test DNA as a template. The products are then mixed with PCR products previously prepared from normal DNA and the mixture is boiled to denature the molecules to single strands. The mixture is then allowed to cool, which allows complementary reannealing to form double strands. In some cases, the shorter molecules, with the deletion, will anneal to longer normal molecules and creating a mismatch, with extra bases on one strand. These molecules will contain a single-stranded loop flanked by complementary double-stranded DNA. The presence of this loop in the heteroduplex causes the molecule to be retarded during migration through a gel. Therefore, heteroduplexes and homoduplexes can be separated.

Diagnosis by PCR may be affected by a number of problems. The reaction itself may not work, because of faulty conditions or an inactive enzyme or because the template is complexed with protein. Another major problem is contamination with exogenous DNA. In a laboratory which frequently uses PCR to diagnose a specific disease, contamination of reagents, pipettes and other laboratory materials with PCR product soon takes place. It is then easy for some 'normal' or 'abnormal' product to contaminate a reaction, and the extreme sensitivity of PCR means that this contaminant will be amplified and a false result will be produced. This is particularly a problem when dealing with single cells since just one cell from the operator's skin will make the result meaningless. Contamination of the reaction with the products of previous reactions can be minimized by reserving one pipette for use before amplification only and by using positive-displacement pipettes with disposable plungers. Introduction of the operator's DNA can only be prevented by taking strict precautions against contamination.

Prenatal diagnosis using linkage analysis

Once the chromosomal location of a gene is known, gene tracking using linked markers can be used for prenatal diagnosis in at-risk pregnancies, even if the specific molecular defect has not been identified. Since the technique is independent of the mutation, it is often preferable to use linkage analysis even when the molecular pathology is known, particularly when the disease may be due to one of a number of mutations. For example, there are over 30 different mutations which can cause β-thalassaemia. These include single-base changes and small and large deletions. It would be pointless to try to identify the specific mutation in any individual family, which could be a time-consuming and expensive exercise, if an affected fetus could be identified by linkage analysis.

For diagnosis by linkage analysis, a linked RFLP or other marker is required. This marker needs to be informative in the individual at risk of transmitting the disease. In the absence of recombination between the linked marker and the disease, it can be deduced that any individual who inherits the RFLP allele which is segregating with the disease in the family will also have inherited that mutation. By studying other members of the family, it can be found which of the RFLP alleles is on the chromosome carrying the gene mutation. This is 'establishing phase'. Whether the individual being tested has inherited that allele, and therefore that chromosome, will indicate his or her risk for the disease. For example, a linked RFLP may have two alleles, A1 and A2 . If the mutation is on the same chromosome as A1, then they are in phase. If an individual inherits A1 from a parent at risk of transmitting the disease, it can be deduced that he or she will also inherit the mutation.

For diagnosis by linkage analysis, the important family members need to be informative for the RFLP being used. This means that, with autosomal recessive diseases, both parents need to be heterozygous. With autosomal dominant diseases, the individual at risk of transmitting the disease needs to be heterozygous and with X-linked diseases the mother needs to be heterozygous. In many cases, a DNA probe will detect several RFLPs, all linked to each other and to the disease. If key family members are not informative for one RFLP, others may be found which are informative. Ideally this initial analysis should be done before the couple embark on a pregnancy so that they can do so in the clear knowledge that prenatal diagnosis is available in their case.

The need to be able to establish phase from other members of the family can place limitations on diagnosis in cases where key family members are unavailable for testing or are dead. Prenatal diagnosis for autosomal recessive disease, for example, is often requested when the only other affected family member is a previous child. If this child is dead, no DNA is available and there are no other affected relatives, it will not be possible to establish phase and diagnosis by linkage analysis will not be possible. In certain cases, use of PCR using the dead child's Guthrie spot may provide the necessary information. In other cases, it may be necessary to identify the specific genetic mutation causing the disease in that family, if that is possible, and to directly analyse the fetal DNA for the mutation. In lethal X-linked disorders, it is known that the carrier must be the mother and it may therefore be possible to establish phase from a healthy son or other male relative.

A further cause of potential error is recombination between the linked marker and the disease. This would cause a change in phase and the disease would then be inherited with the opposite allele (in a two-allele system). The chance of recombination depends directly upon the distance between the marker and the gene. For most linked RFLPs used in prenatal diagnosis, this chance is below 1%, so an error in predicting an affected fetus would occur in less than 1 case in 100. There are examples, however, of diseases with linked markers where recombination has been observed more frequently. The Duchenne muscular dystrophy gene spans a physical distance of about two million bases. Because of the enormous size of this gene and the frequency of recombination events around it, there is recombination between several linked probes and the disease in about 5% of cases. Errors due to recombination may be reduced by analysis using flanking markers, with an RFLP either side of the gene itself. Since the risk of recombination for each is independent, the chance of a double recombination event is very small.

New mutations also present a problem in assessing the risk to an unborn child. In Duchenne muscular dystrophy new mutations account for one-third of new cases. When a woman already has one affected child, the risk to her unborn child clearly depends upon whether or not she is a carrier. This may be difficult to establish. Carrier status is, of course, much more likely if there are other affected males in the family and almost certain if there are two affected sons. A new mutation is more likely if there are a significant number of healthy males in the family. In many cases, it is not possible to be certain whether an individual woman is a carrier or not.

AUTOSOMAL RECESSIVE DISEASES

Requests for prenatal diagnosis of autosomal recessive diseases often come from parents who have one existing affected child and where only a small pedigree is available. Diagnosis then frequently relies on the use of several linked RFLPs. For any individual meiosis to be informative that individual needs to be a heterozygote. If both parents are homozygous for an individual RFLP, then that RFLP will provide no useful information. In Fig. 14.5 both parents are homozygous. It is impossible to say whether the unborn child has inherited the mutation from both parents, as did the affected child, or has inherited the mutation from only one parent, or is completely normal. The only situation in which a full diagnosis can be made in such a small pedigree with only one linked RFLP is where both parents are heterozygotes and the affected child is a homozygote. In this case (Fig. 14.6), analysis of the alleles inherited from the affected child shows that the mutation is inherited with A1 from both parents. If the unborn child is A2 A2, it will be normal, and if A1 A2 it will be a carrier. If both parents are heterozygotes but the affected child is also a heterozygote, as in

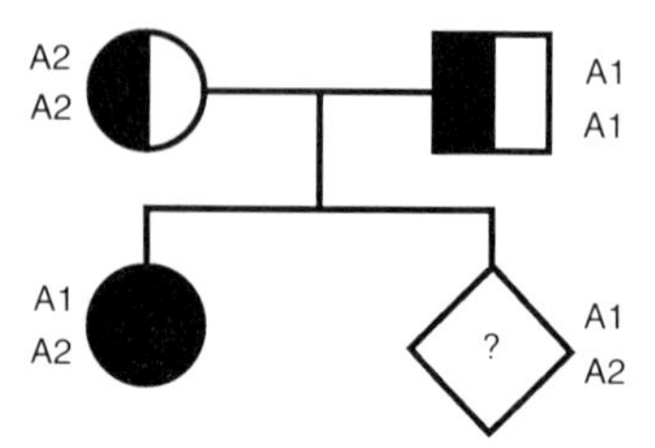

Fig. 14.5 Prenatal diagnosis of an autosomal recessive disease by RFLP analysis. Both of the parents are homozygous. It is not possible to predict the risk for the fetus in this case. (See Appendix 2 for key to symbols.)

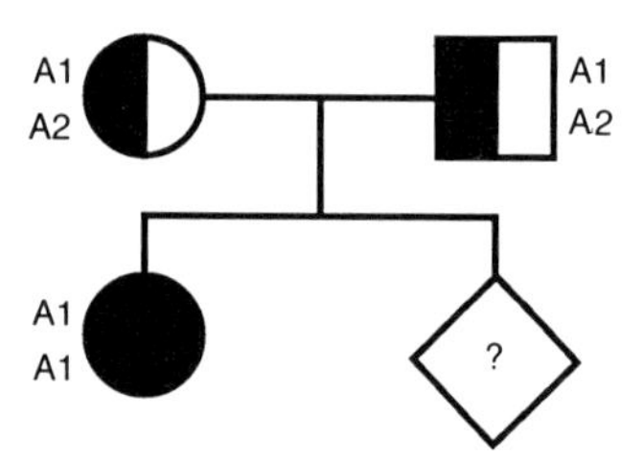

Fig. 14.6 Prenatal diagnosis of an autosomal recessive disease which is fully informative for an RFLP. Both of the parents are heterozygous (A1 A2) and the affected child is homozygous (A1 A1). If the fetus is A2 A2 it will be unaffected, if it is A1 A2 it will be an unaffected carrier and if it is A1 A1 it will be affected. (See Appendix 2 for key to symbols.)

Fig. 14.7, it will be impossible to distinguish an affected fetus from a normal fetus. If the fetus is A1 A2, it could have inherited either both mutated or both normal genes. If the fetus is A1 A1 or A2 A2, then it must have inherited one mutated gene and one normal gene and can be identified as a carrier. Figure 14.8a shows a situation where with only one linked RFLP only a partial diagnosis can be made. The affected child has inherited A2 from the father. Since the fetus must have inherited A2 from the father, it must have at least its father's mutation. The mother is homozygous A1 A1 so it is not possible to say whether the fetus has inherited her normal gene or her mutated gene. With this haplotype it has an equal chance of being either a carrier or affected. If the fetus had been A1 A1, then it could not have inherited its father's mutation and would have an equal chance of being either normal or a carrier. This information allows a so-called 50% diagnosis to be made. Either the fetus is normal or a carrier, which is usually entirely 'good news'; or the fetus is affected or a carrier, in which case the parents might prefer to terminate the pregnancy. In the latter case, a full diagnosis might be made by using oligonucleotide probes or some other method of direct detection of the mutation in the fetus. In the case of the haemoglobinopathies, diagnosis by fetal blood sampling and globin chain synthesis studies might be made. In the example given, the addition of information from a second linked RFLP for

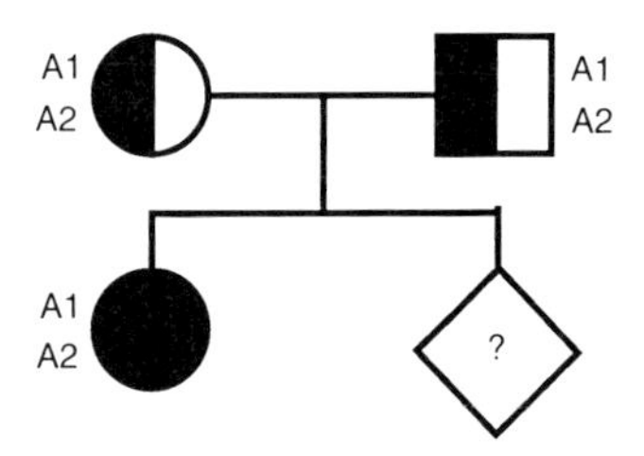

Fig. 14.7 Prenatal diagnosis of an autosomal recessive disease using RFLP analysis. In this example, the pedigree is only partially informative because the affected child has the same haplotype as both of the parents. If the fetus inherits an A1 A2 haplotype, it could be either affected or normal. If it inherits either A1 A1 or A2 A2, it will be a carrier. (See Appendix 2 for key to symbols.)

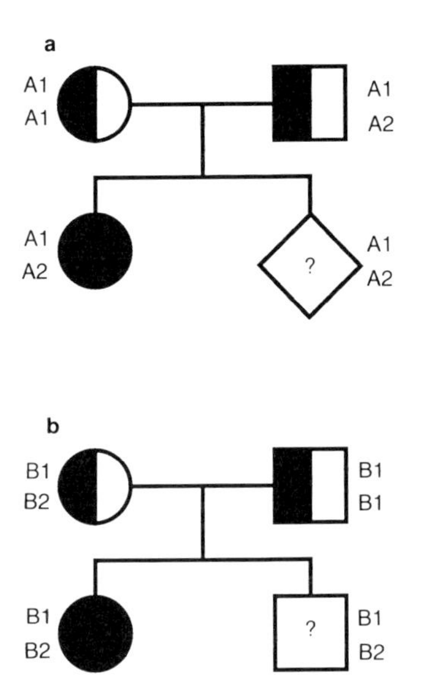

Fig. 14.8 Use of a second RFLP in prenatal diagnosis of an autosomal recessive disease. (a) The affected child indicates that the father's A2 allele is from the affected chromosome. However, it is not possible to tell which of the mother's chromosomes has been inherited. Therefore there is not enough information to decide whether the fetus will be affected or a carrier. (b) Using a second linked RFLP which is informative in the mother allows a complete diagnosis to be made. (See Appendix 2 for key to symbols.)

which the mother is heterozygous makes a full diagnosis possible by linkage analysis (Fig. 14.8b).

When the index child has died and no DNA is available, it is usually necessary to make use of the extended family and more than one RFLP. In Fig. 14.9 the index child has died and DNA is not available from him. Two RFLPs have been used with genotypes A1 A2 and B1 B2. The unaffected child has inherited the B2 allele from each parent. He must also have inherited the A2 allele from his mother. Therefore, we know that in the mother the A2 and B2 alleles are in

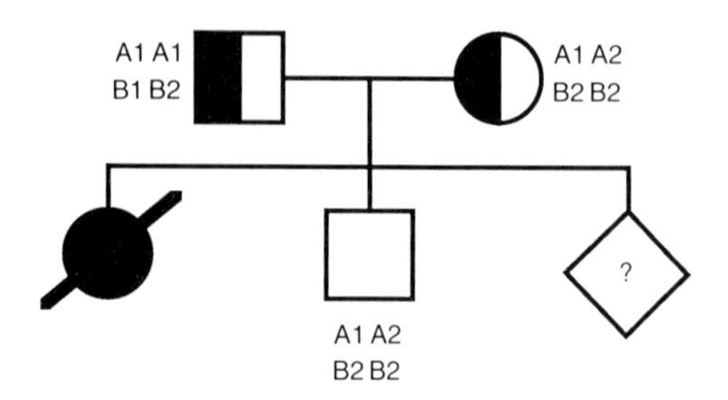

Fig. 14.9 Prenatal diagnosis of an autosomal recessive disease using RFLP analysis. The affected child in this family is dead and DNA is not available. Since phase cannot be established from the parents, both RFLPs are uninformative for the diagnosis of the fetus unless the haplotype is the same as for the unaffected son. (See Appendix 2 for key to symbols.)

phase, i.e. they are on the same chromosome. We do not know whether the unaffected child is completely normal or a carrier. In this case, prenatal diagnosis for the unborn child can only exclude the disease if it has exactly the same genotype as the unaffected son. Figure 14.10 shows the extended family. The mother's brother is a carrier, since he has an affected daughter. This affected child must have inherited A2 B1 from its mother, and with it the disease mutation. The remaining alleles, A2 B2, inherited from its father, must also be in phase with the disease. So we know now that, when considering the unborn child, on the maternal side of the family the disease is in phase with A2 B2. The unaffected son has these alleles from his mother and so must have inherited the mutation from her. Since he is normal, he must have inherited a normal gene from his father together with the alleles A1 B2. It can then be deduced that if the unborn child is A1 B1, A2 B2 it will be affected.

These are deliberately simple examples. Frequently larger pedigrees and larger numbers of linked RFLPs are needed to provide useful information. For example, linkage analysis with eight linked RFLPs can provide full fetal diagnosis of thalassaemia in 92% of Italian

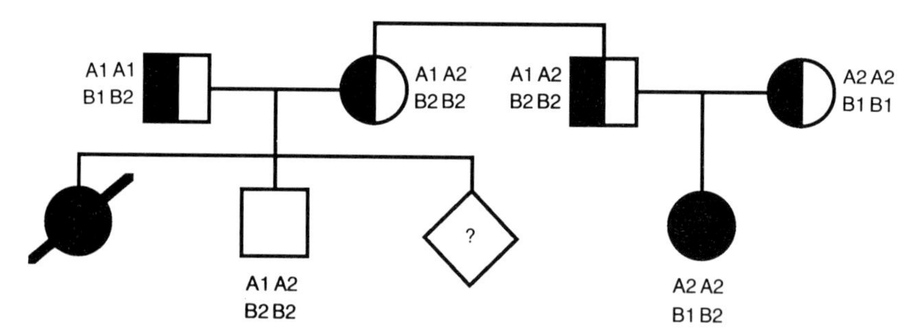

Fig. 14.10 Analysis of an extended family can often be useful in determining the phase. In this example, data from the mother's brother allows phase to be determined so that a diagnosis in the fetus can be made. (See Appendix 2 for key to symbols.)

families and 70% of Cypriot families. It is also possible to examine a pedigree to determine whether or not prenatal diagnosis is indicated. It is possible that a couple will request prenatal diagnosis because their relatives have the disease although they themselves have not yet had an affected child. This is particularly likely in inbred pedigrees. It is often possible in these cases to demonstrate that one of the partners is not a carrier, thus avoiding the risks of intervention for prenatal diagnosis.

AUTOSOMAL DOMINANT DISEASES

With autosomal dominant diseases the affected parent needs to be heterozygous for informative prenatal diagnosis. Other family members are then needed to establish phase. In Fig. 14.11 the father of the unborn child has the disease. He has inherited A2 from his father with the disease mutation and A1 from his normal mother. The disease mutation is therefore in phase with A2. The fetus must inherit A2 from its mother. If it inherits A2 from its father, it will be affected. If it is heterozygous A1 A2, it will be unaffected.

In the case of myotonic dystrophy, recombination presents an additional problem. Recombination between the linked probe and the disease occurs in about 1% of meioses in sperm but in about 4% of meioses in oocytes. This built-in error needs to be taken into consideration when predicting the risk of the disease in the fetus.

The late-onset autosomal dominant Huntington disease also presents a special problem with prenatal diagnosis. Individuals who have an affected parent have a 50% chance of being affected themselves. They will also realize that their offspring, therefore, have a 25% chance of inheriting the disease. Although such potential parents may wish to know the risk to their offspring, they frequently do not wish to know whether they themselves have inherited the disease. In certain cases it is possible to exclude disease in the fetus without generating infor-

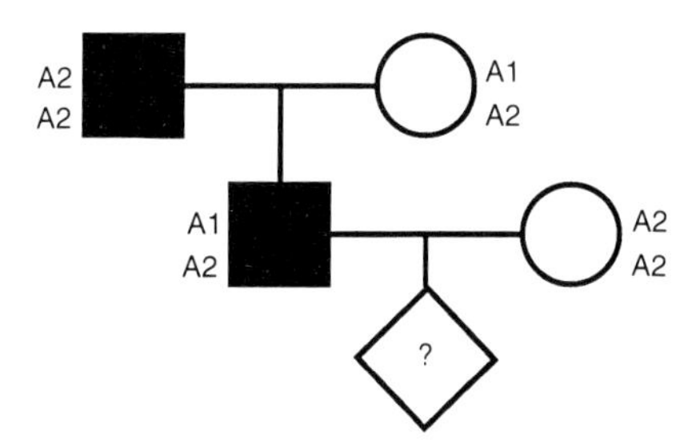

Fig. 14.11 Prenatal diagnosis of an autosomal dominant disease by RFLP analysis. Phase in the father can be established from studying his parents. In this example the disease is in phase with the A2 allele. If the fetus inherits the A2 allele from its father, giving the genotype A2 A2, it will be affected. If it inherits the A1 allele from its father, giving A1 A2, it will be normal. (See Appendix 2 for key to symbols.)

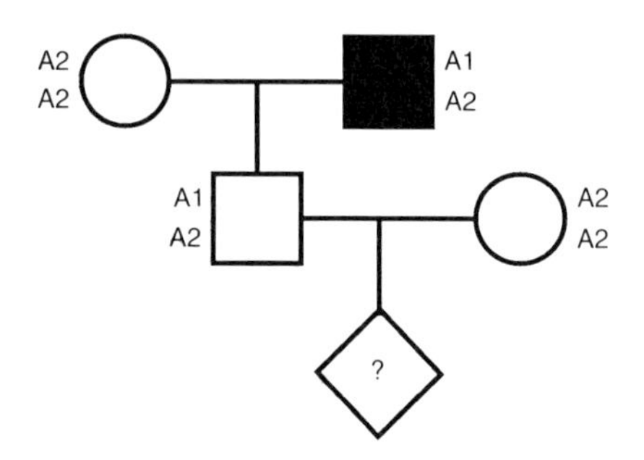

Fig. 14.12 This example demonstrates prenatal exclusion of Huntington disease without generating additional information about the parents' risk of developing the disease. If the fetus is A2 A2, it will have inherited its grandmother's Huntington locus and will therefore not be at risk. If it is A1 A2, it will have inherited a grandpaternal gene. Since the phase is unknown, its risk in this case is 50% and the parents may choose to have the pregnancy terminated. (See Appendix 2 for key to symbols.)

mation about the parents' risk of developing the disease. In Fig. 14.12 the father is heterozygous for the linked allele. His mother is homozygous A2 A2; therefore he must have inherited A2 from his mother and A1 from his father. The fetus must inherit an A2 from its mother. If it also inherits A2 from its father, it can be deduced that it has inherited that chromosome 4, or at least the part of it bearing the Huntington locus, via its father from its grandmother. Since the grandmother did not have Huntington disease, the fetus will be normal. No additional information is generated about the risk to the father since we cannot tell whether the disease is in phase with A1 or A2. If the fetus inherits A1 from its father and therefore has the genotype A1 A2, with the information that we have here it has a 50% risk of inheriting the disease. Unfortunately such prenatal exclusion is not possible in all cases. For example, if the father had been A2 A2, then prenatal exclusion would not have been possible in this case. If the extended family is studied to establish phase, information is often generated about the risk of disease in the parents.

X-LINKED DISEASES

The principal characteristics of an X-linked recessive disease are that only males are affected and that an affected father cannot have an affected son. All of the daughters of an affected man must be carriers. For a linked RFLP to be informative in an X-linked condition, the mother needs to be heterozygous. The father, of course, will be homozygous since he only has one X-chromosome. An additional complication in X-linked diseases is the frequency of new mutations.

The simplest case is seen in Fig. 14.13. The mother is heterozygous A1 A2, the affected child is A2 and therefore the fetus, if it is A1, will be unaffected. If the fetus is A2, it will probably be affected unless the

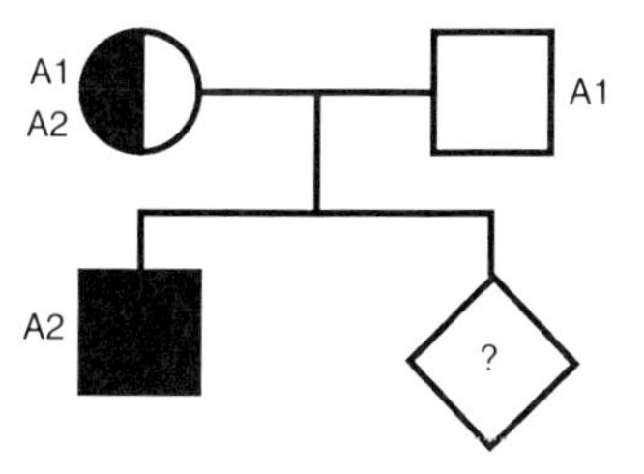

Fig. 14.13 Prenatal diagnosis of an X-linked disease by RFLP analysis. The affected child has inherited the A2 allele from his mother. If the fetus is female, she will be unaffected although she may be a carrier if she inherits the A2 allele. If the fetus is male, however, inheritance of the A2 allele indicates that he will be affected. (See Appendix 2 for key to symbols.)

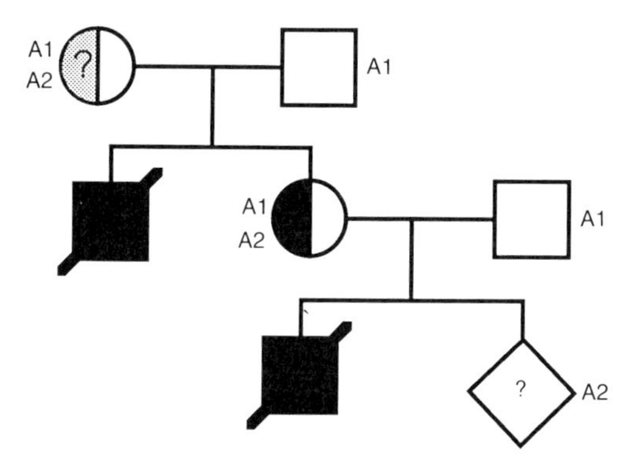

Fig. 14.14 The possibility of new mutations can complicate linkage analysis in X-linked diseases. In this example, the mother has an affected son from whom DNA is not available. Since her father has the A1 allele and is unaffected, it appears that she inherited the mutation from her mother with the A2 allele. Since the mother had an affected brother her affected son was unlikely to represent a new mutation. Using the phase obtained from the mother's father, the unborn child has inherited the affected chromosome. (See Appendix 2 for key to symbols.)

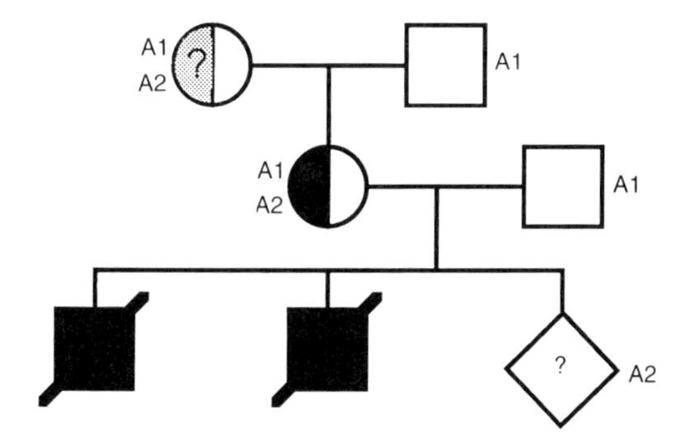

Fig. 14.15 In this example, the mother has had two affected sons from whom DNA is not available. She is an obligate carrier. Since her father has the A1 allele and is unaffected, she inherited the mutation from her mother with the A2 allele. The fetus will therefore be affected. However, the carrier mother could represent a new mutation, which would negate the prediction of phase that had been made. No diagnosis could then be made for her unborn son. (See Appendix 2 for key to symbols.)

affected child represents a new mutation. A larger pedigree may be needed to determine whether or not a mother is a carrier. In Fig. 14.14 the mother has an affected son from whom DNA is not available. This in itself does not confirm that she is a carrier, since the son may have a new mutation. But she has an affected brother, which confirms that she is a carrier. Since her father is A1 and unaffected, it appears that she inherited the mutation from her mother together with A2. Phase can be determined from the mother's father. Even when a mother has had two affected sons and is therefore an obligate carrier, the possibility that she herself represents a new mutation may make establishing phase impossible (Fig. 14.15).

USING THE POLYMERASE CHAIN REACTION IN DIAGNOSTIC LINKAGE ANALYSIS

PCR can also be useful in diagnosis using linkage analysis and has the advantage of speed. The actual mutation causing the disease does not need to have been identified. Sequence data are needed for the region flanking the restriction site which causes the linked RFLP. If it is known from studies in the family that the disease is segregating with a particular RFLP, that RFLP could be detected in a small amount of fetal DNA using PCR. Primers would be designed which produce a fragment spanning the restriction site in question, the area would be amplified and the fragment would then be digested with the appropriate restriction enzyme. If the site is present, two smaller fragments would result; if not, then the fragment size would remain unchanged. Using PCR in this context is faster and less expensive than the conventional method using Southern blotting. This approach has been used for prenatal exclusion of cystic fibrosis in families with one affected child. An

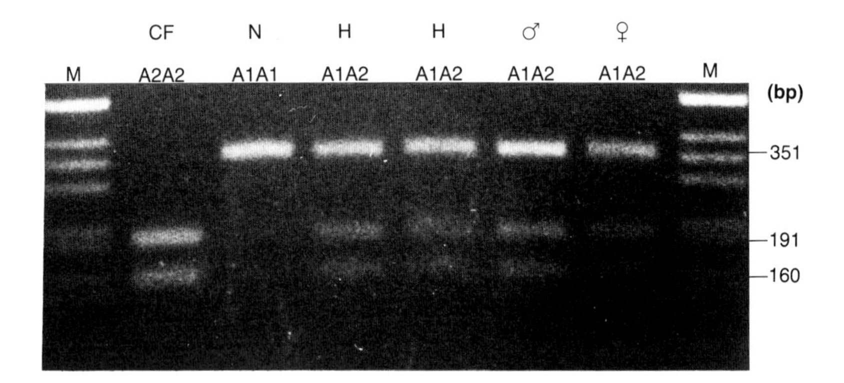

Fig. 14.16 Determination of haplotypes in a cystic fibrosis family by restriction digestion of PCR-amplified products. PCR products amplified from the pCS.7 locus in family members were digested with the restriction enzyme *HhaI*. The resulting fragments were analysed on a 3% agarose gel stained with ethidium bromide. CF = cystic fibrosis, N = normal, H = carrier heterozygote, ♂ and ♀ = parents, M = molecular weight marker.

RFLP with *HhaI* is detected by the probe pCS.7, which is linked to the cystic fibrosis locus. PCR primers have been made so that the product spans this *HhaI* site (Fig. 14.16). Since this particular RFLP is in linkage disequilibrium with the cystic fibrosis locus, most chromosomes carrying the mutation also carry that particular restriction site, although for certain diagnoses the phase needs to be established from other family members. One difficulty with this approach is the sensitivity of some restriction enzymes to reaction conditions. A 'no site' result might be due to failure of the enzyme to cut even though the recognition site is present. One solution to this problem is to hybridize the PCR products with oligonucleotides which detect the polymorphic restriction site. One oligonucleotide is designed to detect the sequence with the restriction site present, whilst the other will detect the same region but with the point substitution which destroys the restriction site. Hybridization of oligonucleotides to PCR products gives more reliable results than hybridization to genomic DNA, since there is a much larger concentration of target DNA in PCR products (Fig. 14.17).

Preimplantation diagnosis

The ability of the PCR to amplify specific regions of the genome from as little as one copy has led to attempts to diagnose single-gene defects in embryos conceived *in vitro*. The intention is that carrier parents would enter an *in vitro* fertilization programme, several embryos would be conceived and only unaffected embryos would be placed into the uterus. Any spare unaffected embryos might be frozen and replaced when a further pregnancy is desired. At first sight this might seem to be an inconvenient, costly, emotionally trying and very inefficient approach to the problem. But it has been argued that when a couple is at high genetic risk, with up to a 75% chance of a termination of pregnancy after diagnosis by chorionic villus biopsy or early amniocentesis, many would prefer preimplantation rather than prenatal diagnosis. The period

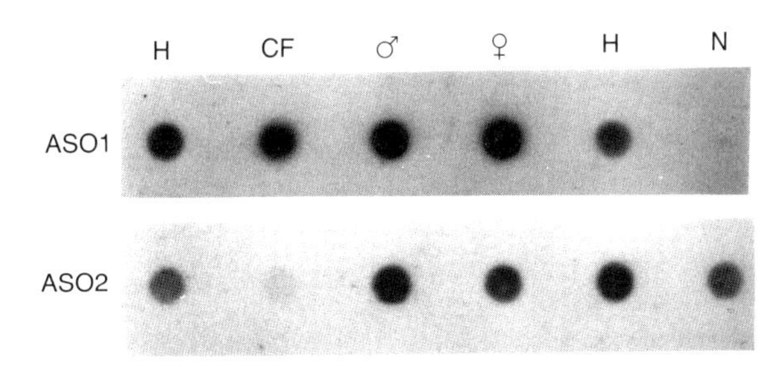

Fig. 14.17 Detection of polymorphic haplotypes using oligonucleotide hybridization (allele-specific oligonucleotides, ASOs). DNA sequences at this locus differ by a single base but, by using optimized hybridization conditions, this can be detected using a pair of oligonucleotides. One of the pair will form a perfect duplex and will remain bound and the other will contain a single-base mismatch and will wash off. (The same family members and PCR products are used as in Fig. 14.16.)

between conception and prenatal diagnosis for these people is one of fear and anxiety. The early pregnancy is often concealed from friends and relatives and the parents themselves experience ambivalence in their bonding to the pregnancy until the results of prenatal tests are known. There are also couples who, for religious or other reasons, are unable to accept termination of an affected pregnancy but will accept *in vitro* fertilization with preimplantation screening.

OBTAINING EMBRYONIC DNA

Both the first and second polar bodies contain genetic information and attempts have been made to make diagnoses using polar body biopsy. Unfortunately, this is not an entirely suitable method for diagnosis since recombination events during meiosis may make it impossible to predict the genotype of the egg from the genotype of the polar body. The first polar body contains a duplicated haploid DNA complement. If the mother were heterozygous, e.g. for a cystic fibrosis mutation, then, provided that no recombination takes place, the polar body and the oocyte will be homozygous. If the first polar body proves to contain the cystic fibrosis mutation, then it can be deduced that the oocyte will be normal. However, if recombination takes place, then both the first polar body and the oocyte will be heterozygous and it will be impossible to predict whether the oocyte or the second polar body will 'inherit' the cystic fibrosis mutation. The second polar body is not formed until after fertilization and this problem could only then be resolved by biopsy of the second polar body as well. A second major disadvantage of polar body biopsy is that only the maternal half of the genome of the embryo can be studied; it would not be possible to know whether the embryo had inherited an abnormal paternal gene.

It is therefore probably better to biopsy the embryo itself. In certain animals, e.g. sheep, the blastocyst reaches a large size, containing several thousand cells before implantation. Quite large numbers of cells can be removed to allow the blastocyst to be sexed before embryo transfer. Unfortunately, this is not the case in the human; however, it has been found that removal of a single cell from an eight-cell embryo does not prevent implantation and normal development. Once a single cell has been removed, PCR may be used to try to identify the presence or absence of a genetic abnormality (Fig. 14.18). At the time of writing, this technique is in its infancy. One of the earliest uses for preimplantation diagnosis was in determination of embryonic sex in women who carry X-linked diseases, e.g. Duchenne muscular dystrophy. PCR primers were designed to amplify a specific region of the Y chromosome. This region is present as a repeating motif throughout the Y chromosome. Although only a single cell was used, there were larger numbers of actual templates for PCR. In theory, male embryos therefore showed a band of the appropriate size and female embryos showed no band (Fig. 14.19).

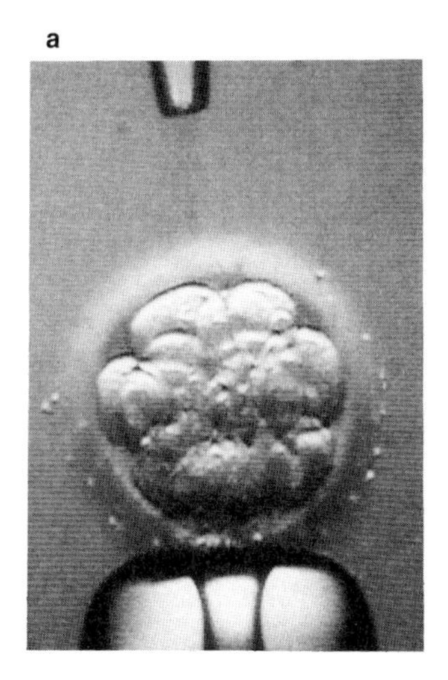

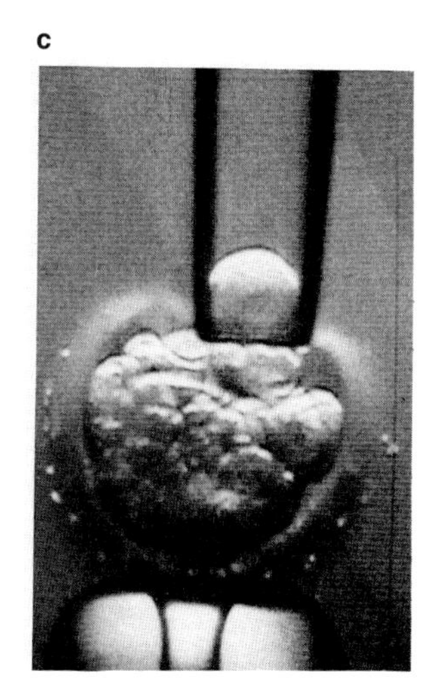

Fig. 14.18 Biopsy of a 12-cell human embryo on day 3. (a) The embryo is immobilized by gentle suction on a flame-polished holding pipette. (b) A fine micropipette is used to drill a hole in the zona pellucida, using a stream of acid Tyrode's solution. (c) Finally a larger micropipette is pushed through the zona to aspirate a single cell. Note that the nucleus of the cell biopsy is visible. Stills from video courtesy of Dr Alan Handyside, Institute of Obstetrics and Gynaecology, London.

There are several difficulties with this approach, which are also applicable to single-locus PCR using single cells. Failure of the reaction to work might cause a male embryo to be incorrectly diagnosed as female. This might occur through enzyme failure or faults in the reagents or template. If the single template cell becomes lost or adherent to the test-tube, the reaction may fail and the embryo will be assumed female. It also appears that some of the cells in an eight-cell embryo may be anucleate. Presumably these cells do not contribute any further to the development of the embryo, which continues normally since the nucleate cells remaining are still totipotential. Biopsy of an anucleate cell from a male embryo would produce no PCR bands and a mistaken diagnosis of female. These problems may be reduced by running a second reaction in the same tube using primers for a locus on another chromosome, but even this cannot guarantee that the primary reaction would have worked if a Y chromosome were present. Contamination of the reaction with foreign DNA may lead to the misdiagnosis of a female embryo as male. This may be avoided by having only female operators set up the reactions. But contamination by DNA may come from other sources, e.g. the pipettes or the reagents themselves. Even commercially supplied reagents may be contaminated with minute quantities of bacterial, phage, plasmid or human DNA. Although this may be of no consequence in general molecular biology work, the presence of two molecules of contaminating DNA is of great significance when there are only a few molecules of template DNA in the reaction. It is clear

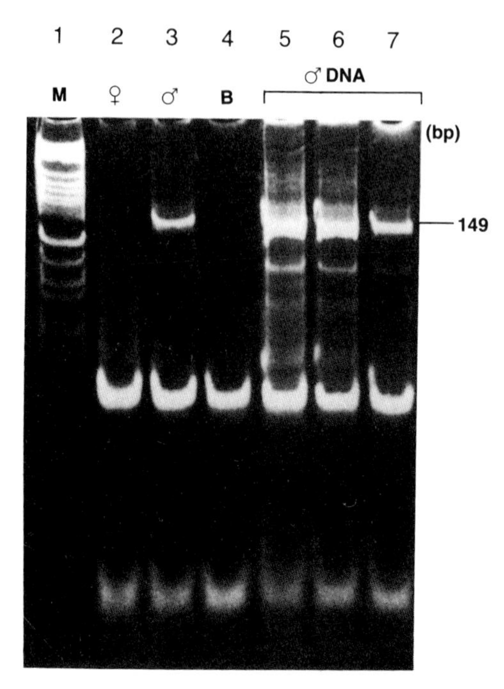

Fig. 14.19 Analysis of PCR products produced by a pair of oligonucleotides designed to amplify a 149 base pair fragment of a 3.4 kb Y chromosome specific repeat sequence, on a 12% polyacrylamide gel stained with ethidium. Lanes (from left to right): (1) DNA size markers, (2) amplification from a single cell biopsied from a female embryo, (3) amplification from a single cell biopsied from a male embryo, (4) control — no DNA, (5) control — male DNA 20 ng, (6) control — male DNA 200 pg, (7) control — male DNA 2 pg. The specific 149 bp PCR product (arrowed) is seen only in lanes containing male DNA. The bands below represent primer dimer formation and single primers. Courtesy of Dr Alan Handyside, Institute of Obstetrics and Gynaecology, London.

that very stringent controls need to be performed alongside diagnosis by PCR.

There are only isolated reports of pregnancy following preimplantation diagnosis for single-gene defects. When amplifying a single locus, there are, of course, only two templates for PCR in a single cell. This reduces the chance that the reaction will work satisfactorily but, whatever the genotype of the embryo, there will be a PCR product if the reaction works. Should the reaction fail, this will be recognized and the embryo will not be considered to be either normal or abnormal.

As improvements in both PCR and *in vitro* fertilization technologies take place, preimplantation diagnosis many become more efficient and gain wider popularity. At present it is applicable to a very small population and it is unlikely that chorionic villus biopsy and amniocentesis will be displaced as the mainstays of prenatal diagnosis.

Screening for genetic defects

A screening test for genetic defects needs to be able to identify all of those with the defect or, if screening for a carrier status, it needs to identify all carriers. That is, it needs to be highly sensitive. The more specific the test is, the fewer normal individuals it incorrectly identifies as positive, but, since secondary investigations are usually done to confirm the result of a screening test, it is sensitivity which is the more important. In Great Britain, genetic screening strategies can be divided into two groups: those applied to neonates shortly after birth and those applied to potential parents before conception or in early pregnancy.

In the first group the most general screen is the examination of the neonate by a paediatrician or obstetrician shortly after birth. This may reveal evidence of chromosomal abnormalities or abnormalities associated with single-gene defects, which can then be confirmed by specific testing, e.g. by undertaking karyotyping. The sensitivity of this form of screening depends upon the skill and alertness of the examining doctor. In Great Britain, there is a specific screening programme for phenylketonuria, using the Guthrie test. Phenylketonuria is a single autosomal recessive gene defect which causes a deficiency of phenylalanine hydroxylase. In the absence of this enzyme, there is a block in the conversion of phenylalanine to tryrosine. The accumulation of phenylalanine in the circulation causes damage to the central nervous and other systems. The Guthrie test uses a phenylalanine-dependent strain of *Bacillus subtilis* as a biological indicator of raised serum phenylalanine. The disease can be controlled by early intervention and dietary modification. A test for hypothyroidism is performed on the same blood sample. Although specific molecular biological tests are becoming available for many single-gene defects, it is unlikely that these will become applicable to any neonatal screening programmes since these conditions are relatively rare in the general population. In the very long term, when there is effective treatment, possibly by gene replacement, it may be appropriate to screen all neonates for a series of single-gene disorders. The neonatal period also presents an opportunity to screen for carrier status in recessive and X-linked conditions. Whether this is more appropriately done in the neonatal period or in adult life will be discussed later.

In general, tests which are offered to the parent before conception or in early pregnancy are only really of value when termination of the pregnancy might be considered if the fetus proved to have inherited the disease. The screening test may be applied to all members of the population or only to a selected high-risk group. For example, all women booking for prenatal care have their haemoglobin, mean cell volume and mean cell haemoglobin measured. The finding of a low mean cell volume and mean cell haemoglobin raises the possibility that the mother is a thalassaemia heterozygote. A similar blood picture is seen in iron-deficiency states but, once an individual has been identified as 'positive', further specific investigations can be carried out to confirm the mother's thalassaemia status. If a thalassaemia trait is confirmed, the father can be similarly investigated and, if the fetus is found to be at risk, invasive prenatal diagnosis can be offered. The new molecular biology techniques have made little impact at this level of screening in this area but are proving invaluable in further investigation of pregnancies screened as positive. Sickle-cell anaemia is an example of a condition where screening is limited to the racial group at risk. However, great care needs to be taken to ensure that the entire population at risk is screened. As immigrants become widely integrated into British society, there is intermarriage between immigrants and the indigenous population and it becomes increasingly difficult to be sure who is at risk.

Screening for trisomy by invasive techniques — amniocentesis or chorionic villus biopsy — is usually limited to women in the high-risk group over 35 years old. It is well established now that, although this approach will reduce the incidence of trisomy in this small group, it makes very little difference to the overall incidence of the condition, since most trisomy babies are born to the much larger number of younger women having babies. The 'triple test' or 'Barts test', in which the simultaneous measurement of maternal serum α-fetoprotein, oestradiol and human chorionic gonadotrophin can give a clearer indication of the risk of trisomy and may prove to be a better method of screening than simple maternal age alone.

With the exception of the haemoglobinopathies, there is at present no screening programme in Great Britain for the more common single-gene defects. The most common of these is cystic fibrosis, with a carrier frequency of between 1 in 20 and 1 in 25. This means that approximately 1 in 500 couples are at risk of having a baby with cystic fibrosis and that, since this is a recessive genetic disease, the risk that their offspring is affected will be 1 in 4. The overall incidence of cystic fibrosis is therefore about 1 in 2000. The average maternity unit will deliver two or more cases each year. At present, screening for cystic fibrosis is only offered to parents who already have an affected child, since they are carriers of the disease. This means that invasive investigations of the fetus can be carried out, and its genotype determined, either by direct identification of the mutation or by using linkage analysis. As discussed before, 70% of cystic fibrosis mutations are due to a single codon deletion, which can be detected using PCR in blood or buccal cells. One possible screening strategy would be to screen all pregnant women in the first trimester for this common deletion. The partners of women who are found to carry the deletion could then be investigated for carrier status. If a carrier father is also found, then the status of the fetus could be determined from DNA obtained at chorionic villus biopsy or amniocentesis, and termination of pregnancy offered in cases where the fetus is affected. Such a strategy could reduce the incidence of cystic fibrosis from 1 in 2000 to 1 in 7000. To further reduce the incidence, the mother would need to be tested for the other common mutations, which would greatly increase the cost of the screening programme.

If screening tests for cystic fibrosis and other single-gene defects become available, the time point at which they should be applied may become a matter for debate. An individual's first contact with health care services is at birth, when all babies are tested for phenylketonuria. In the case of phenylketonuria, screening at birth is essential because treatment needs to be instigated immediately. Screening for carrier status for other single-gene defects at birth would have the advantage that very few individuals would miss inclusion in the programme. Unlike phenylketonuria, however, detection of a cystic fibrosis carrier would have no consequences until the individual wished to begin a family. Health care records are not sufficiently well organized in Great Britain for the result of a screening test at birth to be readily available 20 or more years later. Giving this information to the parents of the neonate might cause them unnecessary psychological distress. Additionally, there may be difficulty in passing the information on to the child when he or she is old enough to understand it, and there is a very real danger that carrier children or young adults could become stigmatized within the community.

Screening for rubella immunity is undertaken in Britain in schoolgirls as they enter secondary school. The principal advantage of the timing of this test is that the few who have not acquired immunity can be immunized before they reach their reproductive years. Screening for carrier status for single-gene defects at this time would seem to be inappropriate for the same reasons as for the screening of neonates. Screening in the first trimester of pregnancy has the advantage that the test is immediately relevant to the pregnancy and only those couples in which both screen positive will be subjected to worry. The disadvantages are that many women miss early prenatal care and some receive no care at all, and that screening once a pregnancy is established does not give the option to consider adoption or preimplantation diagnosis. The ideal point at which to screen may be immediately before conception; however, very few healthy women attend preconception clinics and at present there are not the resources within the health service to cater for a large increase in demand.

Such screening programmes also raise a number of ethical considerations. Frequently calculations are made to determine the cost of the screening programme and the cost of caring for affected individuals if screening is not undertaken. If the cost of screening is less than the cost of caring for affected individuals, then screening is thought to be worth while. This line of argument appeals to financial planners but fails to take into consideration the less easily measured social and psychological burden of the disease on the affected individual and close relatives. At present, such screening programmes also operate on the assumption that an affected pregnancy will be terminated — that it is better not to be born than to be born with a congenital abnormality. Yet, for example, although both physically and mentally handicapped, individuals with Down syndrome are frequently happy people who appear to enjoy life. When investigating a pregnancy at risk for congenital abnormality using invasive techniques, it is often thought that, if termination of the pregnancy would not be acceptable to the couple, then the test should not be undertaken. But it can be argued that knowledge of the abnormality early in pregnancy would allow such couples time to prepare themselves for the birth of an abnormal child.

15: Molecular biology technology in the treatment of genetic disease

At present, the main thrust in the reduction in morbidity and mortality due to genetic disease is in preventing the birth of affected neonates. Preimplantation diagnosis has been successfully applied in only a handful of cases and so this essentially means that the only course of action is the termination of affected pregnancies. In general, with the exception of infections, there are very few diseases which can be cured in the strictest sense of the word and this also applies to genetic diseases.

Currently the major role for molecular biology techniques in the treatment of disease is the commercial production of proteins using cloned genes. The mainstay of treatment for haemophilia is the replacement of the missing factor VIII purified from whole-blood donations. It is becoming clear that haemophiliacs treated in this way are at risk of infection by the human immunodeficiency virus (HIV). Molecular biology technology can be applied to this problem in two ways. The highly sensitive polymerase chain reaction can be used to identify infected blood donations. But, more importantly, 'recombinant' factor VIII synthesized by bacteria containing cloned human genes, without the risk of HIV, is becoming available. Similarly the risk of both HIV and Jacob–Creutzfeld disease in children with hypopituitarism who are treated with growth hormone extracted from cadaver pituitaries will be abolished by the increased availability of 'recombinant' growth hormone. Recombinant insulin is now widely available and has become the treatment of choice for newly diagnosed diabetics, since it reduces the risk of the disadvantageous immune response seen with animal insulins.

In these cases molecular biology techniques are being used to produce pharmacological agents. A more satisfactory approach would be to replace endogenous production of the missing factors. This might be achieved using either of two strategies. Either foreign cells producing the missing factor could be introduced into the host, essentially a form of transplantation, or the faulty gene could be 'replaced' into the host cells themselves.

Therapy by transplantation

One of the best-known attempts to treat a genetic disorder by transfer of normal cells is in the treatment of thalassaemia by bone marrow transplantation. The patient's own bone marrow needs first to be destroyed by irradiation and is then replaced with marrow from a normal donor. If the donor is HLA-compatible, the transplanted marrow will usually then function normally. Although HLA typing is frequently performed using the mixed lymphocyte reaction, typing by restriction analysis is becoming increasingly common. Unfortunately, there is only a 1 in 4 chance that a sibling will have an identical HLA type. The results of non-HLA-matched transplants are disappointing. As well as the risk of bone marrow rejection, there is also the risk of graft-versus-host disease, in which lymphocytes derived from the 'foreign' graft attack other host tissues.

One interesting approach to this problem is to identify the fetus which has inherited thalassaemia and to give intrauterine marrow injections by ultrasound-guided needle during the second trimester. In theory the fetal immune system would be insufficiently mature to mount a rejection response and will become tolerant of the transplanted antigens. Such experiments have so far only been undertaken to a limited extent in animals, but this may come to represent one of the first examples of our ability to treat rather than terminate an affected pregnancy.

Transfer of foreign myoblasts into patients with Duchenne muscular dystrophy may prove to be a successful technique for replacement of missing dystrophin. Muscle fibres are regenerated from these normal myoblasts and should synthesize dystrophin normally. It is possible that they would proliferate, replacing the abnormal dystrophin-negative fibres and returning muscular function to normal.

Gene therapy

The ultimate achievement in the application of molecular biology technology to the treatment of disease would be the correction or replacement of mutated genes. It will be more difficult to remove a defective gene than to add a complementary 'corrected' version. There are a large number of technical problems which need to be addressed before therapy by gene transfer can become a reality. Some form of delivery system needs to be designed which can safely introduce foreign DNA into somatic cells with a reasonable degree of efficiency. It is essential that, once incorporated into the host, transferred genes are expressed in the correct tissues and with the correct regulation. Housekeeping genes, expressed at a relatively low level in most cells at all times, would not need to be regulated as carefully as some genes which require precise control. In certain cases where the genetic defect is in a gene whose expression does not need to be carefully controlled, it is possible that a low level of expression of transferred genes might correct the biochemical abnormalities. Examples of conditions where this might be applicable are Lesch–Nyhan disease, purine nucleoside phosphorylase deficiency and blood clotting factor deficiencies. In the case of clotting factor deficiencies, experiments have shown that it might prove to be possible to introduce the gene into skin fibroblasts, which would then produce the protein in sufficient amounts to ameliorate the disease. There is likely to be more difficulty in the case of genes requiring precise control. For example, although a great deal is known about the structure and function of haemoglobin genes, not enough is known about their regulation for it to be possible to transfer normal globin genes in the near future. Over-expression of β-globin from genes transferred into a patient with β-thalassaemia could result in the clinical phenotype of α-thalassaemia in that patient and his or her offspring.

A way will need to be found in which some control might be exerted over transferred genes. If all of the controlling elements associated with a gene could be identified and transferred together with the gene itself, then it might be found that the inserted gene came under the same regulatory control as its natural homologue and was expressed only in the correct cells at the correct times. The identification of enhancer sequences may prove to be of value. These regulatory sequences may lie quite some distance from the gene itself. If some way could be found to link transferred genes with tissue-specific enhancer sequences, it might be possible to provide a degree of cellular specificity.

There is also a problem with transferring the normal genes into the correct target cells. If foreign genes were introduced into totipotential embryonic cells, or by some vector which recognized all cell types, then the inserted gene might enter the germ line as well as somatic cells. The engineered sequences would then be passed on from generation to generation. From an ethical point of view, transfer of genes into the somatic cells of one individual is little different from organ transplantation, but modification of the germ line would probably represent an ethically unacceptable use for molecular biological technology.

Gene transfer

GENE TRANSFER BY CALCIUM PRECIPITATION

The technology for introducing foreign genes into cells in culture and into fertilized mouse eggs is now well established. Cells growing in culture can be transformed with DNA when precipitated with calcium. A small number of cells will take up the foreign DNA and incorporate it at apparently random sites within its genome. Various techniques have been developed to detect those cells which have taken up the

DNA from the general cell population. These involve cotransforming the cells with a selectable marker in a way similar to the selectable markers on plasmid vectors which allow recombinants to be differentiated from non-recombinants. One such marker is the prokaryotic neomycin resistance gene. Eukaryotic cells are sensitive to neomycin, which is toxic to ribosomes. Hence only transformed cells will grow in a medium containing neomycin. Interestingly, it was originally thought that the transforming gene and the selectable marker gene would need to be ligated together before being introduced into the cell to ensure that they are incorporated together in the host genome. This proved not to be the case; it was found that the two genes needed only to be in the same calcium DNA precipitate to be incorporated together.

GENE TRANSFER BY MICROINJECTION

A second technique for introducing foreign genes into host cells is by micromanipulation and injection. It is possible to inject foreign DNA into recently fertilized mouse eggs, usually into the male pronucleus. The embryo is then either directly implanted into a foster-mother or implanted after *in vitro* development to the blastocyst stage. A few of these pregnancies will go to term and it will then be found that, in a small proportion of these, foreign DNA has become incorporated into the host genome. It appears that any number of foreign DNA molecules may be incorporated and that they are often arranged in head-to-tail arrays of varying copy number, usually with no chromosome specificity. The foreign DNA is then found in all cell types in the new-born mouse, including the germ line, and can be passed on from generation to generation. A strain of mouse containing foreign DNA in this way is known as a 'transgenic' strain. One of the earliest and best-known examples of the use of this technique was the creation of a strain of transgenic mice containing the rat growth hormone gene linked to the 5′ regulatory sequences from the mouse metallothionine gene. The idea was that the inserted growth hormone gene could be induced by treating the mouse with metals. These mice express the inserted gene and grow more quickly and to a larger size than normal mice. Transgenic mice have become an invaluable tool in the study of gene function and regulation. It may be possible in the future to detect genetic defects in human embryos prior to implantation and to use similar techniques to correct the defect. However, this approach would introduce engineered material into the germ line, which may not be morally acceptable.

Viral vectors

SV40

The techniques of calcium precipitation and microinjection are applicable only to cells in culture or to very early embryos. They may require selectable markers and are very inefficient. Currently, the method for gene transfer which appears to hold the greatest promise is the use of viruses as vectors. The mode of infection of any virus is to identify a specific cell type, latch on to receptors on the cell wall and inject DNA or RNA into the cell. In many cases, virally coded genetic material then becomes incorporated into the host genome and may remain there permanently. These properties of viruses make them obvious candidates for mediating the transfer of foreign genes into animal cells. One of the first viruses to be examined for its potential in gene transfer was the SV40 virus. SV40 is a monkey DNA tumour virus. Its DNA is a closed circle bearing 'early' and 'late' genes. The early genes are responsible for the initiation of viral DNA replication and the late genes for the synthesis of the proteins needed to package the virus. In certain cell types — non-permissive cells — viral replication cannot take place and the early genes instead cause malignant transformation. It is possible to replace the late genes of SV40 with foreign genes. The virus will transfect monkey cells but, since it cannot make the late protein, remains in the cell either as a plasmid-like extrachromosomal molecule or incorporated into the host genome. Viral replication can be made to take place by cotransfecting a helper virus which replaces the function of the late genes and allows synthesis of the packaging proteins. SV40-derived vectors have been used to introduce the gene for the enzyme hypoxanthine-guanine phosphoribosyltransferase into monkey kidney cells in culture. This is the gene coding for the deficient enzyme in the Lesch–Nyhan syndrome. The transfected cells appear to produce the missing enzyme. It was work with SV40 which originally demonstrated the existence of 'enhancer sequences'. These short DNA sequences, when cloned on a plasmid with another cloned gene, greatly increase its expression. Their action seems to be unrelated to their 5′ to 3′ orientation on the plasmid and they may work by providing a region of DNA free from bound protein, to which RNA polymerases can bind and begin translation.

RETROVIRUSES

The vectors receiving the most interest for the transfer of foreign genes into eukaryotic cells at present are retrovirus vectors. Retroviruses contain RNA as their genetic material. Once the virus has entered the cell, a DNA copy of its RNA is made by viral reverse transcriptase. The cDNA then becomes incorporated into the host genome, where it is expressed to produce both the new viral genomic RNA and all of the proteins needed to form new virions. Virions are then released from the host cell by budding, without lysis of the cell itself. The viral genome consists of long terminal repeat sequences, which contain the genes needed for incorporation into the host cell, and the *gag*, *pol* and *env* genes, which code for the virion core proteins, the reverse transcriptase enzyme and the envelope proteins respectively. Experiments have been performed in which the *gag*, *pol* and *env* genes have been replaced by a selectable marker, e.g. a neomycin-resistance gene, and a cloning site for a foreign gene. This construct was then transfected into a fibroblast cell line. Simultaneously the cells were infected with a helper virus which produced the missing *gag*, *pol* and *env* proteins. The virions released into the culture medium then possessed the ability to infect new host cells but could not replicate new virus. Experiments with mice have shown that these retrovirus vectors can be used to transfer genes into stem cells from the bone marrow. When reinjected into an irradiated mouse, these cells become established in the spleen, where they express the transferred genes.

At present, the only human disease which has been successfully treated using gene therapy is adenosine deaminase (ADA) deficiency. ADA deficiency is a bone marrow disorder which inhibits the growth and production of T and B lymphocytes, leading to severe combined immunodeficiency. This could previously only be cured by HLA-matched bone marrow transplant. ADA deficiency proves to be a good target for gene therapy, as a range of gene expression from 5 to 5000% of normal appears to cure the disease. In the first human ADA-deficient patients to be treated, a retroviral vector containing the ADA gene and an SV40 promoter was used. Cells were harvested from the bone marrow and infected with the construct. After several cycles of growth in cuture, the cells were returned to the patient. It has been found that, although only 1% of cells were gene-converted, the healthy cells soon outgrow the others. Preliminary trials of gene therapy for ADA deficiency are producing promising results.

Other approaches to gene therapy

Other approaches to gene therapy include the targeted modification of gene mutations and the use of suppressor tRNA genes. Target modification means the use of exogenous DNA molecules which contain the correct gene sequence to directly replace the mutated sequence. If the replacement sequence is otherwise identical to the genomic sequence, it may be possible to incorporate it at the correct site by making use of natural recombination events. Suppressor tRNA may be used to treat diseases due to point mutations. Point mutations will either place an inappropriate stop codon within the gene or cause the incorporation of the wrong amino acid in the growing protein chain. Transfer RNA genes can be engineered so that the tRNA that they produce will recognize the altered codon but incorporate the ‘correct’ amino acid. If a suppressor tRNA gene were to be transferred into a cell carrying such a genetic defect, its tRNA would compete with the endogenous tRNA for the codon and at least some of the protein product would be normal. What effect this might have on the expression of other genes in the cell is not known. Both of these techniques are not so much in their infancy as at the blastocyst stage of development, but, if successful, both have the advantage that they represent a specific site-directed approach to gene therapy at the genetic level.

16: DNA fingerprinting

Single-copy, unique-sequence DNA probes will detect only two sites in the human genome, one on each chromosome. Since the mean rate of heterozygosity in the human genome is about 1 per 1000 base pairs, many single probes will not detect any polymorphisms. Many RFLPs are dimorphic, i.e. there are only two alleles. With a two-allele system, with the maximum degree of heterozygosity, one-quarter of the population will be homozygous for one allele, one-quarter will be homozygous for the other allele and a half will be heterozygous. It would be impossible to use such an RFLP to identify DNA from a specific individual. If, however, several hundred single-locus probes were used, each of which detected an RFLP, then a specific allele pattern would arise which would be specific to any individual. This is the basis of DNA fingerprinting, but, rather than using any single-locus probes, one probe is used which detects a repeat unit dispersed throughout the genome.

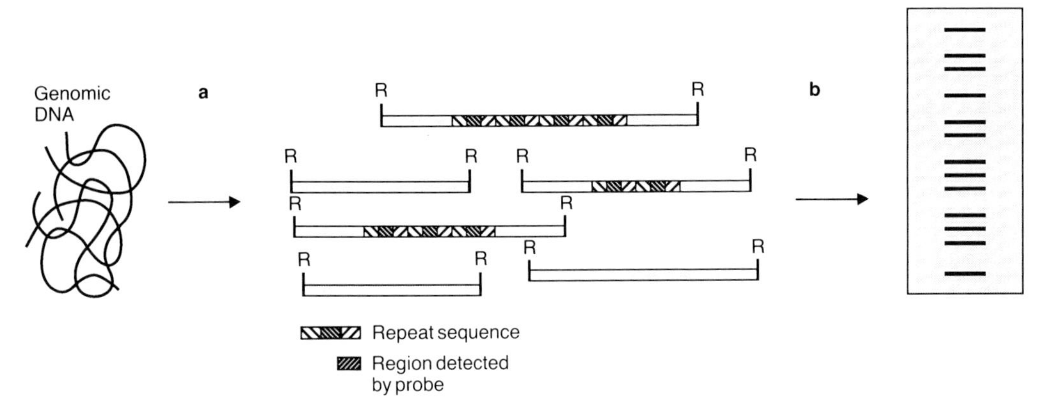

Fig. 16.1 DNA fingerprinting using minisatellite repeat probes. (a) High molecular weight DNA is cut with a restriction enzyme. Distributed throughout the genome are 'minisatellite' regions consisting of a variable number of repeats of a short sequence. Between individuals the number of repeats in any specific fragment is different. (b) The various restriction fragments are separated by electrophoresis and Southern blotted. The blot is hybridized to a probe which detects the central core of the repeating unit. After autoradiography the fragments containing the minisatellite regions are seen as a ladder of bands. Although individuals may share one or more bands, the overall banding pattern is unique to one individual (except in the case of identical twins).

The variability in restriction fragment lengths detected by these probes is further increased because the repeat unit itself contains regions of variable length.

The human genome contains many dispersed loci which consist of tandem repeats of a short sequence and which are known as 'minisatellite regions' (Fig. 16.1). The number of repeats at each minisatellite is variable and the number of repeats at any particular locus is variable between individuals. This high degree of variability is thought to have arisen from unequal recombination at these sites during meiosis or from slippage during DNA replication. A probe was constructed which detects these minisatellite regions by ligating together several copies of the core repeat sequence. Almost all of the work concerning hypervariable minisatellite regions has been done by Professor Alec Jeffreys and his group in Leicester and these probes have come to be known as Jeffreys probes.

If DNA from an individual is digested with a frequent-cutter restriction enzyme, Southern blotted and hybridized to a Jeffreys probe, the resulting autoradiograph will show a large number of bands. It will be seen that any individual will have inherited half of the bands from each parent, but the degree of polymorphism is so high that it appears that the banding pattern for each individual is unique. Hence the term DNA fingerprint (Fig. 16.2). More recently, probes have been introduced which detect unique sequences which are very highly polymorphic. These are also based on variable numbers of tandem repeat sequences (VNTR) but may be specific for an individual chromosome. These can be used in the same way as the original Jeffreys probe but are easier to read since each individual will usually have only two bands rather than the large number seen with the original probes (Fig. 16.3).

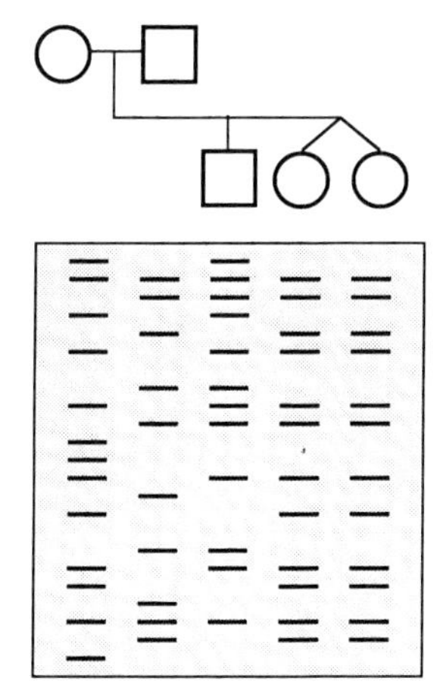

Fig. 16.2 DNA fingerprinting using minisatellite repeat probes within a single family. The mother and father each have a unique banding pattern. Their son has inherited some of his mother's bands and some of his father's bands to produce his own unique pattern. Their daughters have also inherited a mixture of parental bands but, since they are identical twins, they share an identical banding pattern.

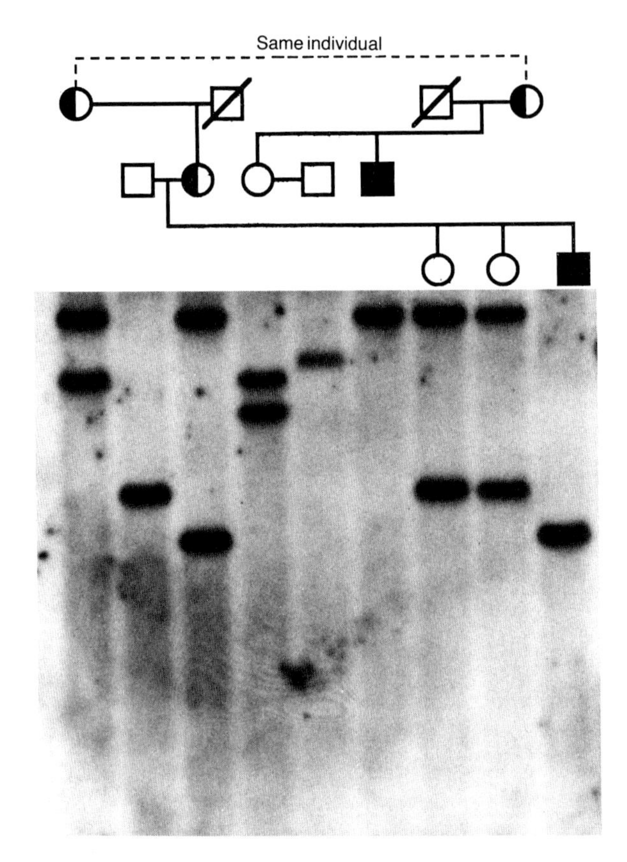

Fig. 16.3 Use of an X-linked VNTR probe in a linkage analysis study on a large family.

DNA fingerprinting has found particular application to forensic medicine. DNA can be extracted from blood or tissue found at the scene of a crime and matched to that of individual suspects. In rape cases DNA obtained from semen can be used for matching. Although each individual spermatozoon has only a haploid DNA complement, and would only show half of the bands, a mixture of spermatozoa produces a DNA fingerprint identical to any somatic cell from the same individual. Some difficulties have arisen in the use of DNA fingerprinting to identify specific individuals. To show that an individual's DNA matches DNA found at the scene of a crime, it has to be shown that their banding patterns are identical. Unfortunately, variations in the electrical field at electrophoresis are common. These cause DNA in different lanes to run slightly differently. This may be obvious, and in non-forensic cases the experimenter will compensate for it when reading the autoradiograph. In forensic cases, such adjustments may be unacceptable since what appears to be a different banding pattern may genuinely be due to wrongful identification of the suspect. Unfortunately, the amount of DNA extracted from material found at the scene of the crime may be insufficient to allow several repeat Southern blots to be performed.

DNA fingerprinting has also been used to confirm maternity or paternity in cases of dispute about immigration or child maintenance. Since any individual will inherit half of the bands on his or her DNA fingerprint from each parent, it is a simple matter to demonstrate that a particular child is the offspring of a particular couple. There are fewer technical difficulties with this application of DNA fingerprinting since larger amounts of DNA are usually available for multiple Southern blotting. Furthermore, whereas in a criminal case, where the identification must be proved without any doubt, in a civil case the court is willing to accept a high level of probability.

Although single-locus VNTRs show a high degree of polymorphism within the general population, they cannot be used to discriminate monozygous from dizygous twins. Although identical twins will display identical alleles, since only one allele is inherited from each parent there is a 25% chance that non-identical siblings will also display the same alleles. In this situation, a multilocus probe will give the necessary information.

There is one interesting clinical application for DNA fingerprinting. A complete hydatidiform mole has a normal diploid chromosome number but all of its chromosomes are inherited from the father. A partial mole is triploid, with two sets of paternal chromosomes and one set of maternal chromosomes. The differentiation between complete and partial mole is usually made histologically by identification of both molar and normal products of conception in partial moles, or molar tissue only in complete moles. Occasionally, however, this differentiation can be difficult. The diagnosis can then be made by DNA fingerprinting. A complete mole will inherit only paternal alleles whilst a partial mole will inherit both paternal and maternal alleles.

17: Molecular biology techniques in microbiology

Bacterial typing

Bacterial infections are readily diagnosed and the infecting organism can usually be cultured without difficulty. Molecular biology techniques have been applied in modern microbiology to identify bacterial types and subtypes. Whereas with digestion of human genomic DNA with almost any restriction enzyme a smear of DNA will be produced after electrophoresis, the reduced complexity of bacterial DNA means that selected restriction digests will produce a clearly definable characteristic banding pattern. Comparison of the banding pattern produced by the test organisms with those from known strains and subtypes will allow the type of the test organism to be determined. Alternatively, a Southern blot can be made and probed either with a probe specific for a particular type or a probe common to many or all bacterial types. For example, a probe which detects the gene for ribosomal RNA should detect a band in all cases. The type is then assigned based on knowledge of the restriction size detected by that probe in various preparations of known type. It is also possible to isolate bacterial plasmids from unknown organisms in the same way that plasmids are isolated from host bacteria during cloning experiments (see Chapter 10). Identification of plasmid types using similar techniques may help in the identification of their host organism.

Bacterial identification using molecular biology techniques is principally of interest to the microbiologist. However, the application of molecular biology to the detection and identification of *Toxoplasma*, viruses and their subtypes is of increasing interest in obstetrics and gynaecology.

Toxoplasma infection in pregnancy

Transplacental transmission of *Toxoplasma gondii* from an acutely infected mother in early pregnancy results in congenital *Toxoplasma* syndrome. This may cause death *in utero* or severe neurological damage. Infection later in pregnancy may be asymptomatic at birth but neurological damage may become evident later in childhood. Approximately 20% of women with *Toxoplasma* infection in pregnancy will give birth to a transplacentally infected infant. Clearly, to offer termination of pregnancy to all women who acquire primary infection in the first or second trimester of pregnancy would cause the unnecessary abortion of many uninfected fetuses. Prenatal diagnosis permits treatment of the fetus as well as providing information that is valuable when making decisions about termination of the pregnancy. The most efficacious regimen for treatment of an infected fetus is the combination of pyrimethamine with sulphonamides. This antibiotic combination is potentially hazardous and is usually reserved for cases in which fetal infection has been definitely proved.

The standard methods of prenatal diagnosis are detection of IgM antibodies in fetal blood and culture of amniotic fluid and fetal blood, both in mice and in fibroblast culture. IgM testing has not proved to have the specificity and sensitivity required for confirmatory testing. Fibroblast cell culture may provide a diagnosis in a few days but is effective in less than 50% of cases. Isolation by mouse inoculation is far more sensitive, but is more expensive and complicated and may take several weeks to provide the diagnosis. Although a delay of up to 6 weeks between initial testing and termination of pregnancy may be less significant from a technical view in the first trimester than in the second, the psychological consequences to the mother are considerable at both stages of pregnancy.

A method using PCR has been established for the detection of *Toxoplasma gondii* in cells pelleted from amniocentesis. PCR primers have been designed to amplify a 193 base pair region of the *Toxoplasma* B1 gene. The function of this gene is unknown but it is repeated 35 times in the *Toxoplasma* genome and is therefore a better target template than a single-copy sequence. PCR detection of *Toxoplasma gondii* has proved to be a promising method for prenatal diagnosis of congenital infection from amniotic fluid. False positives have been found, particularly when testing is performed in laboratories is which large amounts of *Toxoplasma* DNA is handled. These can be reduced by careful handling of the PCR reagents and separation of the diagnostic facilities from the general laboratory. Occasional persistent false negatives have been found. This may be due to absence or mutation of the B1 gene in certain strains of *Toxoplasma*. Nevertheless, it appears that PCR may meet the need for a rapid and definitive diagnosis in infected mothers, applicable across a wide range of gestational ages.

Human genital tract papillomavirus infection

There is increasing evidence for the role of human papillomaviruses in the aetiology of cervical intraepithelial neoplasia and carcinoma of the cervix. There are over 40 different types of human papillomaviruses, but only types 10, 11, 16, 18, 31, 33 and 35 are associated with genital lesions. Type 6 has such a high degree of genetic homology to type 11 that they are now considered to be one type — 11. It appears that human papillomavirus type 11 is associated with benign condylomas and low-grade cervical dysplasias, whereas types 16 and 18 are frequently found in association with carcinoma of the cervix. It is thought that human papillomaviruses act upon the dividing and differentiating cells of the cervical epithelial squamocolumnar junction, causing a loss of the normal control of proliferation, which leads to dysplasia and ultimately to carcinoma. As part of the early replication cycle of the virus, its genome becomes incorporated into the host genome. It is at this stage that rapid cellular proliferation occurs. Viral capsid protein synthesis is a late phenomenon in the viral life cycle and is usually detected only in more mature cells in the superficial layer of the epithelium. Unless virus is actually being produced by the cells, the presence of the virus can only be detected by showing the presence of its genome. Even when virus is released by the cells, detecting its presence can be difficult.

There is no reliable serological test for human papillomavirus infection, and culturing the virus *in vitro* is extremely difficult and certainly insufficiently reliable for use either in epidemiological studies or clinical practice. There are characteristic findings on cytology which suggest human papillomavirus infection, in particular koilocytes, known as 'balloon' or 'halo' cells. However, in many cases it is impossible to differentiate between human papillomavirus infection and dyskaryosis associated with cervical dysplasia. Similarly, there are histological appearances associated with human papillomavirus infection, including epithelial atypia, basal cell hyperplasia and hyperkeratosis, and the presence of koilocytes, but most of these features are also features of cervical dysplasia and even the more specific features, such as keratinization and the presence of koilocytes, may be due to infection by some other virus. Electron microscopy and immunocytochemistry are more reliable in identification of the presence of viral particles, although they will not differentiate subtypes nor will they indicate the presence of viral DNA incorporated into the host cell's genome.

Molecular biology techniques are the only methods which will identify the presence of viral DNA, whether complete virions are present or not, and that will allow identification of the viral subtype. The earliest work used the dot blot technique. This is a technique similar to Southern blotting in which DNA is extracted from cervical biopsies and blotted on to hybridization filters. The filters are then hybridized with a probe from viral DNA of a known type. In theory, with this technique, if viral DNA is present the dot blot will hybridize to the probe, producing a signal on autoradiography, whereas a dot blot which does not contain viral DNA will not produce a signal. In practice, there is a considerable amount of non-specific hybridization of the probe to any DNA present in the dot blot. If the viral DNA represents only a small proportion of the total DNA present, it will not produce a signal which is significantly brighter than the non-specific background signal.

In situ hybridization of specific probes to histological sections of cervical biopsies is a much more reliable method for detecting the presence of human papillomavirus DNA. Not only is the method more sensitive but it also allows the tissue and cellular localization of the viral DNA to be identified. In cells which are releasing virions, hybridization will be seen both in the nucleus and in the cytoplasm, whereas, in those cells in which the viral DNA is present only in the host genome, hybridization will only be seen in the nucleus. Use of type-specific probes can differentiate one viral subtype from another. The technique is, however, technically difficult, time-consuming and expensive and does not lend itself well to be used in the volume that would be needed for an epidemiological study or routine clinical practice.

PCR is rapidly becoming the technique of choice for the identification of human papillomavirus DNA within cytological and histological samples. It is a rapid technique which requires little operator skill or time, can be automated and can process many samples at once. The results can usually be read from an ethidium-stained agarose or polyacrylamide gel, without the need for Southern blotting or the use of hazardous radiolabelled reagents. Oligonucleotide primers have been designed that will amplify regions of viral DNA specific to each human papillomavirus subtype, each of which produces different-sized amplification products. The sensitivity of PCR is such that it can detect a single human papillomavirus DNA molecule in 10 000 human cells. This is 10^5-fold more sensitive than blotting techniques, which are unable to demonstrate reliably less than one viral molecule per human cell. Its specificity can be confirmed by the detection of an amplified band of a size consistent with the predicted size and the absence of bands of other sizes. If need be, the amplified PCR product can be blotted and hybridized to a type-specific probe, since there is then no shortage of template DNA. Since the original amplification reaction requires so little template, a single biopsy can be subjected to simultaneous detection of several viral subtypes and, unlike any of the other techniques, the results can be made available within hours rather than days.

As with the application of PCR to other areas, contamination of reactions with PCR product from previous reactions will lead to false positive results and the greatest care needs to be taken to ensure that this does not occur.

Human immunodeficiency virus

The human immunodeficiency virus (HIV) is a retrovirus, i.e. it contains RNA rather than DNA. One of the earliest events in infection is the formation of a cDNA copy of the RNA by a reverse transcriptase, followed by integration of this cDNA into the genome of host cells. The principal sites of integration of viral-coded DNA are T lymphocytes, macrophages and the microglial cells of the brain. HIV infection is becoming increasingly important in obstetrics and to a lesser extent in gynaecology in Great Britain, and is already of major importance in many African countries. Perinatal transmission of HIV from infected mother to fetus or neonate is associated with a high risk of infection and mortality in the children. Various rates of infection in neonates born to infected mothers have been reported, ranging from below 20% to over 70%. The standard technique for identification of HIV infection is the demonstration of anti-HIV antibodies in the host circulation. However, after infection with HIV and incorporation of viral-coded DNA into the host genome, it appears that there may be a period of 'latent infection', characterized by low viral expression and absence of a serological response. In adults, this period may be only a few weeks but can be up to 1 year. Serological diagnosis in neonates is particularly unreliable. Maternal IgG antibodies may cross the placenta and persist in the neonatal circulation for up to 15 months. Hence, a neonate may be serologically positive although actually uninfected. Conversely, the fetal and neonatal immune response to HIV infection is often delayed or defective. IgM antibodies have not been found to be reliable markers in neonates. A neonate with negative HIV serology may still be infected.

As with human papillomavirus, the most reliable indicator of the presence of HIV is the demonstration of the presence of virally coded genetic material. Early attempts to demonstrate HIV sequences in Southern blots, using DNA extracted from circulating leucocytes, produced disappointing results, presumably because the quantity of virally coded DNA within the total DNA was very small and could not be easily detected above background hybridization. *In situ* hybridization, using both DNA and RNA probes, has proved to be more successful. Interestingly, in some cases in which only the antisense RNA probe was used, it was possible to demonstrate the presence of viral DNA but not viral RNA. This may have been because in these individuals RNA was present at levels too low for detection, but it may have been that these were cases of early, 'latent' infection, in which active expression of the viral genes had not yet begun.

PCR is again the most sensitive method for detection of HIV DNA. Several pairs of oligonucleotide primers have been designed to amplify the *pol* and *gag* regions of the HIV genome. The use of just one pair of primers is associated with a significant false negative rate. This may reflect genetic variations amongst viral isolates or it may be due to more trivial technical differences. Use of three different sets of primers has been reported to eliminate the problem of false negatives. As with its uses in other situations, PCR to detect HIV is often associated with non-specific amplification of unrelated sequences. These, however, will not usually be of the expected size, and confirmation of authentic PCR products can be made by Southern blotting and hybridization to a probe designed from the region internal to the two PCR primers.

Using PCR to demonstrate the presence of HIV DNA in neonates at risk allows the tests to be performed using very small amounts of blood, collected from the neonate a few days after birth. This avoids the need to take cord blood at birth, which may contain HIV-infected mononuclear cells of maternal origin. Early studies have shown that HIV DNA can be demonstrated in up to 50% of seronegative children born to infected mothers. Whether demonstration of the presence of DNA is associated with progression to viral replication and the development of the acquired immune deficiency syndrome has yet to be shown.

Similar experiments in seronegative high-risk adults — those who have infected partners — have also demonstrated a high rate of the presence of viral DNA. This should be borne in mind by the surgeon who intends to operate on such a patient, and until the more sensitive molecular biology techniques become widely available seronegativity should not necessarily by accepted as proof that an individual is not infected with HIV.

Detection of other viruses

The techniques described for the detection of human papillomavirus and HIV can be applied to the detection of any other virus. PCR has been used to demonstrate the presence of hepatitis B DNA sequences in seronegative at-risk individuals. Single point mutations in the genome of this virus may significantly alter its antigenicity and complicate serological testing. PCR will still amplify sequences that carry mutations within the amplified fragment. PCR has also been used to demonstrate hepatitis B infection in neonates and to show that the virus is present in breast milk. Dot blot hybridization, *in situ* hybridization and PCR have all been used to detect the presence of cytomegalovirus (CMV) in biological specimens (Fig. 17.1). CMV serology is positive in up to 60% of blood donations for transfusion. Such blood should not be given to

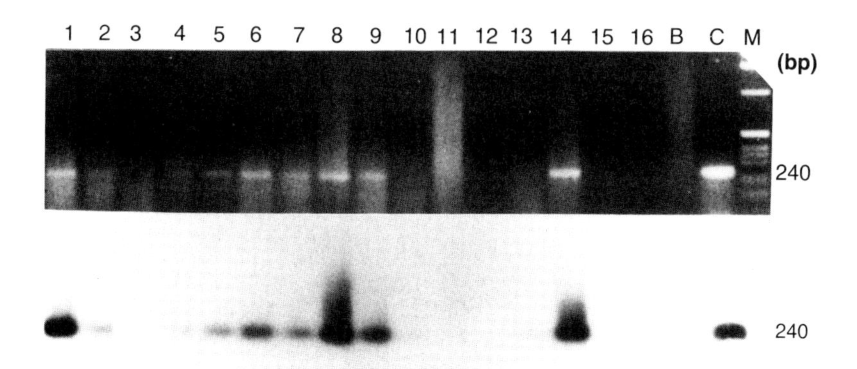

Fig. 17.1 Use of the polymerase chain reaction to detect human cytomegalovirus in urine samples. PCR products were analysed on 3% agarose gels stained with ethidium bromide (top). The gel was then blotted and hybridized with an internal oligonucleotide probe to ensure that the PCR products specifically derived from cytomegalovirus DNA (bottom). Track B contains water blank; track C contains positive control DNA; track M contains molecular weight markers.

immunocompromised recipients. Using PCR it appears that CMV DNA may be demonstrated in about 5% of seronegative donations. Attempts have also been made to detect CMV DNA in chorionic villus biopsies in women thought to have a primary infection in early pregnancy. In these isolated reports, CMV was not detected and the neonate was subsequently found not to be excreting the virus. A similar approach might be taken to identify rubella DNA in the chorion in women suspected of having a primary infection in the first trimester of pregnancy.

Appendices

1: Glossary

Acrocentric: when the centromere is towards one end of the chromosome.
Adenine (A): a purine base of DNA and RNA that pairs with thymine.
Allele: alternative forms of a gene or restriction fragment found at the same locus on homologous chromosomes.
Anaphase: the stage of cell division when the chromosomes move from the equatorial plate to the opposite sides of the cell drawn by the spindle.
Androgenone: egg containing only paternal or sperm-derived chromosomes.
Aneuploid: a chromosome number which is not an exact multiple of the haploid number, i.e. $2n-1$ or $2n+1$ where n is the haploid number of the chromosomes.
Autoradiography: detection of radioactively labelled molecules on X-ray film.
Autosome: all chromosomes apart from the sex chromosomes.
Bacteriophage: a virus which infects bacteria.
Barr body: the persistent mass of the material of the inactivated X chromosome in cells of normal females. Named after Murray Barr, who, with E.G. Bertram, first described sexual dimorphism in somatic cells.
Base pair (bp): a pair of complementary bases of DNA, one on each strand, e.g. A and T.
Bivalents: chromosomes when paired up during cell division.
Blastocyst: the mammalian conceptus in the post-morula stage, consisting of the trophoblast and an inner cell mass.
'Blunt' ends: double-stranded DNA with flush ends produced by cleavage with a restriction enzyme.
'CAT' box: refers to a conserved, non-coding DNA sequence of unknown function about 70–80 bases upstream from the start of transcription.
cDNA: double-stranded DNA complementary to an mRNA.
cDNA library: a set of cDNA clones that contain a representative complement of the expressed sequences in the tissue under study.
Centimorgan (cM): unit used to measure map distance; 1 cM is equivalent to 1% recombination.
Centriole: one of two organelles that radiate out the spindles during cell division.
Centromere: the pinched-in part of the chromosome.
Chiasma: the configuration of crossing-over; plu. chiasmata.
Chromatids: at cell division the chromosome divides into two strands or chromatids pinched in at the centromere.
Chromatin: the nucleoprotein fibres from which chromosomes are made.
Chromomeres: bead-like structures visible on the chromosome during pachytene.
Chromosome: thread-like bodies visible at metaphase made up of DNA and protein.
Clone: all cells derived from a single cell by repeated mitosis and all having the same genetic constitution. DNA cloning is the production of many identical copies of a defined DNA fragment.
Codon: a sequence of three bases (triplet) in RNA or DNA that will code for a specific amino acid.
Colony: a growth of bacteria on a culture plate all arising from the same cell and therefore identical in genetic make-up.
Concordant: a twin pair in which both exhibit the same trait.

Cos site: cohesive ends of a phage DNA molecule.

Cosmid: plasmid DNA packaged *in vitro* into a phage, used as a vector in recombinant DNA technology.

Crossing-over: exchange of genetic material whilst chromosomes are paired at cell division.

Cytosine: a pyrimidine base of DNA and RNA that pairs with guanine.

Deletion: the loss of a DNA segment or part of a chromosome.

Denaturation: disruption of duplex DNA into single strands; change of protein into an inactive form.

Deoxyribose: sugar moiety of DNA, an aldopentose, formula $CH_2.OH.(CHOH)_2.CH_2.CHO$.

Diakinesis: the stage of meiosis I in which the nucleolus and nuclear membrane disappear and the spindles form.

Diploid: the normal genetic status of all cells apart from sex cells; in humans 46 chromosomes.

Diplotene: the stage of meiosis I during which paired, bivalent chromosomes repel each other.

Dispermy: fertilization of a duplicated egg nucleus, or egg and polar body, by two sperm.

Dizygotic: twins from two separately fertilized ova.

DNA: deoxyribonucleic acid, the nucleic acid of chromosomes, containing phosphate, sugar and bases. This macromolecule contains all the hereditary information of the cell.

DNA polymerase I: an enzyme which catalyses the synthesis of double-stranded DNA from single-stranded DNA. Also contains both a 5′ to 3′ and a 3′ to 5′ exonucleolytic activity.

DNA probe: a piece of DNA that can be labelled and hybridized to its complementary sequence.

DNase I: degrades DNA by introducing nicks.

Dominant: genetic expression of a disease in individuals that are heterozygous.

Dot blot: a nylon filter with DNA dotted directly on to it without running a gel before transfer.

Duplication: presence of a segment of a chromosome or DNA in duplicate.

Enhancer sequence: short DNA sequences, which when found or placed next to a coding gene greatly increase its expression.

Eukaryote: an organism whose cells have a true nucleus bounded by a nuclear membrane and exhibit mitosis.

Exon: the segment(s) of a gene that are found in a mature mRNA product and therefore code for the protein.

Expression: synthesis of mRNA from a gene.

F1 generation: first offspring of a genetic cross.

F2 generation: first set of offspring from an F1 generation.

Five-prime (5′) end: the end of a DNA or RNA strand with a free 5′ phosphate group.

Frame shift: mutations which change the reading-frame in which codons are translated into protein.

Gamete: a sex or germ cell containing half the normal chromosome complement.

Gene: the part of the DNA that codes for a specific protein.

Genome: the full complement of DNA, either haploid or diploid.

Genomic library: a set of DNA clones that contain a representative complement of the genome under study.

Genotype: the genetic constitution of an individual.

Giemsa staining: the standard staining technique for the banding of chromosomes at metaphase.

Guanine: a purine base of DNA and RNA that pairs with cytosine.

Gynogenone: egg containing only maternal or egg-derived chromosomes.

Haploid: sex cells or gametes that contain one set or half the normal somatic number of chromosomes; in man 23 chromosomes.

Haplotype: a group of alleles from closely linked loci, usually inherited as a unit.

Heritability: a statistical measurement of the amount that a trait is genetically determined.
Heteroduplex: hybrid duplex formed between two different DNA molecules which are not completely complementary.
Heterozygous: an individual or cell that possesses two different alleles at one locus on a pair of homologous chromosomes.
Homologous: when referring to chromosomes the identical pair; when referring to alleles, of the same locus on a pair of homologous chromosomes.
Homology: level of similarity between DNA, RNA or protein sequences.
Homozygous: an individual or cell that possesses two identical alleles at one locus on a pair of homologous chromosomes.
Housekeeping genes: genes that provide basic functions and are expressed in all cells.
Hybridization: pairing of an RNA and a DNA strand or two DNA strands together.
Hydatidiform mole: a condition resulting from deterioration of circulation of the chorionic villi in a pathogenic ovum, marked by trophoblastic proliferation and cystic cavitation of the avascular stroma of the villi, which come to resemble grape-like cysts. A partial mole has in addition remnants of the fetus and placenta present.
Informative: a restriction fragment length polymorphism that gives different bands in individuals within a family so that linkage analysis can be carried out.
Insertion: the addition of a DNA segment or part of a chromosome.
***In situ* hybridization**: a method of directly visualizing the hybridization of DNA or RNA probes on tissue sections or chromosome metaphase spreads.
Interphase: the stage between two successive cell divisions.
Intron: region(s) of the gene sequence that are not found in the mature mRNA product and therefore do not code for the protein product of the gene.
Inversion: a chromosomal aberration in which part of a chromosome is reversed end to end.
Isochromosome: an abnormal chromosome in which one if the arms is duplicated and the other missing.
Kilobase (kb): a unit of 1000 bases in DNA or RNA.
Kinetochore: a structure at the centromere to which the spindles attach.
Klenow: the large fragment of DNA polymerase I which has only the polymerase activity.
Law of segregation: the separation of characters represented by alleles at meiosis so that each gamete contains only one character.
Leptotene: the stage of meiosis I in which the chromosomes are thread-like in shape.
Ligase: an enzyme used to join DNA molecules.
Linkage: genes that are linked that have their loci within a measurable distance from each other on the same chromosome.
Locus: the position of a gene on a chromosome.
LOD score: logarithm of the odds score — a measure of the likelihood of two loci being within a measurable distance of each other.
Map distance: a measure of the distance between linked gene loci expressed in centimorgans (cM).
Meiosis: the type of cell division of gametogenesis, resulting in half the number of somatic chromosomes so that each gamete is haploid.
Messenger RNA (mRNA): is created directly from the DNA in the nucleus. It transfers the information as a single strand into the cytoplasm and acts as a template for polypeptide synthesis.
Metacentric: when the centromere is in the middle of the chromosome.
Metaphase: the stage in cell division when the chromosomes are lined up on the equatorial plate and the nuclear membrane is no longer present.
Methylase: an enzyme used to methylate DNA.
Minisatellite region: area on the chromosome that contains a high density of repeat DNA.
Mitosis: the type of cell division that occurs in somatic cells.

Monosomy: a condition where one of a chromosome pair is missing.
Monozygotic: twins arising from a single fertilized ovum.
Mosaic: an individual or tissue derived from a single zygote with at least two cell lines differing in genotype and as a consequence often the phenotype too.
Multifactorial: a phenotype determined by many factors both genetic and environmental, each with additional effects.
Mutation: a change in the genetic material, either of a single gene or in the structure or number of the chromosomes. Mutations in the gametes are inherited; those in somatic cells are not.
Non-disjunction: the failure of two members of a chromosome pair to disjoin during anaphase so that both pass to the same daughter cell.
Northern blot: a technique for transferring RNA fragments from an agarose gel to a nylon filter on which they can be hybridized to complementary DNA or riboprobes.
Nucleotide: nucleic acid is made up of many nucleotides each containing a nitrogenous base, a pentose sugar and a phosphate group.
Nucleus: the intracellular organelle found in eukaryotes that contains the genetic material.
Oligonucleotide: a short single strand of a few nucleotides.
Open reading-frame: a stretch of DNA sequence between two stop codons.
Pachytene: main stage of chromosome thickening in meiosis I.
p and q arms: nomenclature of the two arms of a chromosome; the p arm is drawn as the upper, shorter arm with the q arm as the lower, longer arm.
Phage: abbreviation for bacteriophage, which is a virus which infects bacteria.
Phage lambda (λ): a phage cloning vector.
Phenotype: the physical, biochemical and physiological appearance of an individual resulting from the interaction of its genotype and the environment in which the individual develops.
Plaque: clear area on plated bacterial culture due to lysis by phage.
Plasmid: a small, circular DNA molecule capable of self-replication autonomously within a bacterium, used as vectors in recombinant DNA technology.
Poly-A synthetase: enzyme that adds the poly-A tails to the mRNA.
Poly-A tails: sequence of polyadenylic acid attached to the 3′ end of most mRNA strands.
Polylinker: artificially introduced region of a vector which contains a series of restriction sites (not present in the vector) which facilitate cloning.
Polymerase chain reaction (PCR): a method of *in vitro* amplification of DNA.
Polyploid: any multiple of the haploid number of chromosomes, e.g. 3n, triploid, where n is the haploid number of chromosomes.
Primer: a short oligonucleotide that acts as a starting-point on template DNA or RNA for a polymerase to create a double strand.
Progeny: offspring or descendants of a mating.
Prokaryote: an organism without a true nucleus, the nuclear material being scattered throughout the cell, and which reproduces by cell division.
Promoter sequence: recognition sequence for binding of RNA polymerase.
Prophase: the visible stage of cell division when the chromosomes are thickened.
Pseudoautosomal region: region at the top of the p arm of X and Y chromosome which behaves like an autosome during cell division with cross-overs occurring between the X and the Y chromosomes.
Pseudogene: DNA sequence homologous with a known gene that is functionless. Probably derived by mutation of an ancestral active gene.
Purine: $C_5H_4N_4$. A white crystalline organic base, related to uric acid. Adenine and guanine are typical purines.
Pyrimidine: $C_4H_4N_2$. An organic base consisting of a heterocyclic six-membered ring. Uracil, thymine and cytosine are typical pyrimidines.
Recessive: genetic expression of a disease in individuals that are homozygous for a mutation with the heterozygotes being unaffected carriers.
Recombination: the formation of new combinations of linked genes by crossing-over between their loci.
Recombination fraction (θ): frequency of recombination.

Reduction division: the first meiotic division; at this stage the chromosome number per cell is reduced from haploid to diploid.
Restriction enzyme (restriction endonuclease): an enzyme which cleaves DNA at specific sites.
Restriction fragment: DNA fragment produced by a restriction enzyme.
Restriction fragment length polymorphism (RFLP): polymorphism arising from the presence or absence of a particular restriction enzyme site.
Restriction map: a linear arrangement of restriction sites on a piece of DNA.
Restriction site: the set of specific bases at which a restriction enzyme cleaves.
Reverse transcriptase: an enzyme which catalyses the synthesis of DNA from an RNA template; used to synthesize cDNA from mRNA.
Riboprobe: an RNA probe.
Ribose: sugar moiety of RNA, a pentose monosaccharide sugar, formula $C_5H_{10}O_5$.
Ribosomal RNA (rRNA): a component of the ribosomes, functions as a non-specific site for polypeptide synthesis.
Ribosome: minute spherical structures in the cytoplasm, rich in rRNA.
Ring chromosome: a structurally abnormal chromosome in which the end of each arm has been deleted and the broken arms have fused into a ring.
RNA: ribonucleic acid, the nucleic acid formed upon a DNA template, containing ribose instead of deoxyribose and uracil instead of thymine.
RNA polymerase: an enzyme which catalyses the synthesis of RNA from DNA in transcription.
RNase: enzyme which breaks down RNA.
Semiconservative replication: with reference to the method by which DNA replicates. The double helix unwinds and two new strands are made against the two unwound strands.
Sense strand: a strand of DNA that contains the information for the protein product. The antisense strand is reversed and is complementary to the sense strand.
Sex chromosomes: the chromosomes that confer sex on a species, X and Y chromosomes in humans.
S1 nuclease: an enzyme which degrades single-stranded DNA.
Somatic: all cells that are not haploid sex cells.
Southern blot: a technique for transferring DNA fragments from an agarose gel to a nylon filter, developed by E.M. Southern in 1975.
S-phase: phase of a cell cycle in which the DNA is duplicated ready for cell division.
Spindle: microtubules that radiate from the centriole which aid the organization of the chromosomes on the equatorial plate at metaphase and their segregation at anaphase.
Splicing: the removal of introns and the joining of exons during transcription of mature mRNA.
'Sticky ends': double-stranded DNA with a 3′ or 5′ overhang produced by cleavage with a restriction enzyme.
Substitution: the replacement of a base with another, changing the DNA but not altering the number of bases present.
Synapsis: close pairing of homologous chromosomes in zygotene of meiosis I.
***Taq* polymerase**: thermostable DNA polymerase isolated from the thermophilic bacterium *Thermus aquaticus*.
'TATA' box: a conserved, non-coding, promoter sequence about 30 bp upstream from the start of transcription.
Telocentric: when the centromere is at the end of the chromosome.
Telophase: the stage of cell division when the divided chromosomes have been separated and surrounded by new nuclear membranes.
Template: the single strand of DNA or RNA to be copied.
Terminal transferase: an enzyme which catalyses the addition of nucleotides to 3′ ends of DNA.
Three-prime (3′) end: the end of a DNA or RNA strand with a free 3′ hydroxyl group.
Thymine (T): a pyrimidine base of DNA that pairs with adenine.

T_4 polynucleotide kinase: phosphorylates the 5′ end of a DNA molecule using the γ phosphate from a nucleotide triphosphate, e.g. ATP.

Transcription: the process by which genetic information is transmitted from DNA in the chromosomes to messenger RNA.

Transfection: acquisition of new genetic material in eukaryotes by the incorporation of added DNA.

Transfer RNA (tRNA): small RNA molecule which transfers activated amino acids from the cytoplasm to messenger RNA.

Transformation: in prokaryotes, the acquisition of new genetic material by the incorporation of added DNA. In eukaryotes the conversion of normal cells to malignant cells in culture.

Transgenics: introduction of a gene into the genome of a foreign host.

Translation: the process by which the information in the messenger RNA is translated into protein synthesis.

Translocation: the transfer of a part of a chromosome to another non-homologous chromosome. If parts are exchanged between chromosomes the translocation is said to be reciprocal and balanced if no material is lost.

Trisomy: a condition where there are three chromosomes of a given pair.

Uracil (U): a pyrimidine base of RNA that replaces thymine and codes with adenine.

Vector: a plasmid, phage or cosmid into which foreign DNA can be inserted for cloning.

X-inactivation: inactivation of one of the two X chromosomes in female somatic cells.

Zooblot: a blot containing DNA from selected animal species.

Zygote: the fertilized ovum.

Zygotene: the synaptic stage of meiosis I.

2: Standard symbols used in pedigree diagrams

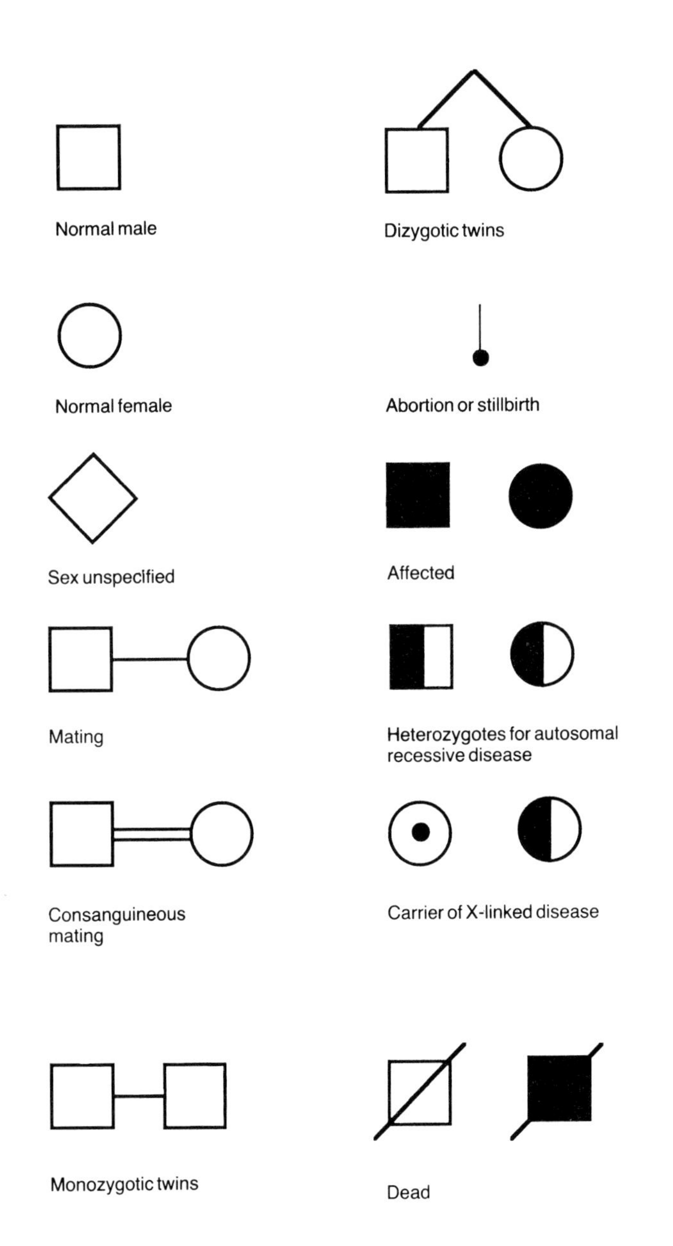

3: Further reading

Innis, M.A., Gelfand, D.H., Sninsky, J.J. & White, T.J. (1990) *PCR Protocols: a Guide to Methods and Applications*. Academic Press, New York.

McKusick, V.A. (1990) *Mendelian Inheritance in Man: Catalogues of Autosomal Dominant, Autosomal Recessive and X-linked Phenotypes*, 9th edn. Johns Hopkins University Press, Baltimore.

Practical Approach Series, The. IRL Press, Oxford.

Ott, J. (1985) *Analysis of Human Genetic Linkage*. Johns Hopkins University Press, Baltimore.

Sambrooke, J., Fritsch, E.F. & Maniatis, T. (1989) *Molecular Cloning: a Laboratory Manual*, 2nd edn. Coldspring Harbor Laboratory Press, New York.

Thompson, J.S. & Thompson, M.W. (1986) *Genetics in Medicine*, 4th edn. W.B. Saunders Company, Philadelphia.

Weatherall, D.J. (1991) *The New Genetics and Clinical Practice*, 3rd edn. Oxford University Press, Oxford.

Index